Lobbying, Political Uncertainty and Policy Outcomes

Sebastian Koehler

Lobbying, Political Uncertainty and Policy Outcomes

palgrave
macmillan

Sebastian Koehler
Department of Politics and Public Administration
University of Konstanz
Konstanz, Germany

ISBN 978-3-319-97054-7   ISBN 978-3-319-97055-4   (eBook)
https://doi.org/10.1007/978-3-319-97055-4

Library of Congress Control Number: 2018949046

Cover pattern: © Melisa Hasan

This Palgrave Macmillan imprint is published by the registered company Springer Nature Switzerland AG
The registered company address is: Gewerbestrasse 11, 6330 Cham, Switzerland

For Maria Catarina and Patrícia

PREFACE

This book fills a gap by bridging several literatures on lobbying which flourish in relative isolation. It is based on my dissertation which I submitted to the University of Mannheim in 2014. In particular, formal models of lobbying and predominantly empirical approaches seldom go hand in hand. The book demonstrates how the current disconnect between formal models of lobbying and empirical studies can be cured. It will be useful for researchers on both sides of the aisle. I hope that we will see more collaboration between proponents of the different approaches in the field in the future.

The process of writing the dissertation was long and arduous. Revising it for publication was a very different journey. I have changed a substantial amount of things. New experiences and many discussions with colleagues have changed my views and improved the book in numerous ways. I am grateful for the advice and comments. To mention just a few of my colleagues from whose openness and kindness I have benefited (in no particular order): Patrícia Calca (ISCTE-IUL Lisbon), Rebecca Morton (New York University), Torun Dewan, Stephane Wolton, Raphael Hortala-Vallve, Brett Mayer, Thomas Leeper, Alexandra Cirone, Ryan Jablonski (all London School of Economics and Political Science), Konstantinos Matakos (King's College London), Thomas König, Thomas Bräuninger, Thomas Gschwend, Hans-Peter Grüner (all University of Mannheim), Patrick Bayer (University of Glasgow), Susumu Shikano (University of Konstanz), Randall Stone (University of Rochester), Andreas Dür (University of Salzburg), Christophe Crombez (University of Leiden and Stanford). Parts of the book have been discussed in the PSPE

Colloquium at LSE, the CDSS Colloquium at the Graduate School of Economic and Social Sciences (University of Mannheim) and various conferences. I thank the participants of these workshops for valuable comments. I would also like to thank the anonymous reviewer for helpful suggestions.

The book would not have been possible without the support of my wonderful wife and the many smiles which I have received from my lovely daughter. I love you both!

Konstanz, Germany   Sebastian Koehler

CONTENTS

VARIABLES AND ABBREVIATIONS

access	Relative access to the Bundesrat
AS	General abbreviation for a Committee in the Bundestag
AWO	Bundesverband der Arbeiterwohlfahrt e.V.
BDA	Bundesvereinigung der Deutschen Arbeitgeberverbände
BDI	Bundesverband der Deutschen Industrie e.V.
BR	Bundesrat
BRaccess	Variable coding access to the Bundesrat
BT	Bundestag
BTaccess	Variable coding access to the Bundestag
ccost	Variable coding lobbying costs in a multiplicative way
C_D	Variable coding intelligence costs
CDU	Christlich Demokratische Union
centdist	Distance to the center of the bargaining range
C_m	Variable coding messaging costs
comdist	Distance to the compromising decision-maker
CSU	Christlich Soziale Union
DGB	Deutscher Gewerkschaftsbund
EU	European Union
FDP	Freie Demokratische Partei
GRÜ	Partei Die Grünen
IGM	Industriegewerkschaft Metall
majf	Dummy variable coding majority status of a bill
nLawChange	Variable coding the number of Laws changed by a bill
nPar	Variable coding the number of paragraphs in a bill
ÖTV	Gewerkschaft Öffentliche Dienste, Transport und Verkehr
SPD	Sozialdemokratische Partei Deutschlands

LIST OF FIGURES

LIST OF TABLES

CHAPTER 1

Introduction

Abstract The introduction defines the research question why interest groups choose a particular communication strategy. I discuss why a new approach to the question is necessary, highlighting the need for a strategic approach. The research design is sketched. I discuss how a tight combination of a formal model and a rigorous empirical analysis helps to shed new light on an old question.

Keywords Lobbying · Strategic decision-making · Communication strategy

Lobbying is one of the most fundamental political activities in representative democracies. It is crucial in the shaping and enactment of public policies. Despite, or maybe because of this, lobbying is highly controversial. The heated debates in the public often operate on murky ground, with little or circumstantial evidence. Before jumping to normative conclusions about lobbying, it is advisable to take one-step back and ask what lobbyists do and which effects lobbying produces.

Politicians decide on public policies in an inherently uncertain world, where it is impossible to fully anticipate the effects of policies. In this world, plagued by fundamental uncertainty, one of the main activities of interest groups is to provide information to policymakers because they have better knowledge about specific effects and aspects of public policies. A manager of a company, for example, is better informed about the cost structure of production and is therefore better able to assess how many people will have to be laid off in response to a tax reform or stricter environmental regu-

© The Author(s) 2019   1
S. Koehler, *Lobbying, Political Uncertainty and Policy Outcomes*,
https://doi.org/10.1007/978-3-319-97055-4_1

lation. This asymmetry drives demand for information by politicians who seek to draft better public policies (Esterling 2004; Truman 1971). Matters become complicated because interest groups are self-interested political actors, pushing for their own agenda. The manager may want to exaggerate the consequences of a tax reform to get a more beneficial outcome for the company. When will interest groups share their information truthfully with policymakers and when will they try to misrepresent the information? The answer to this question depends on the communication strategy of interest groups.

This book[1] contributes both theoretically and empirically to the ongoing debate of how interest groups choose a lobbying strategy. To advance the debate, I focus on a clearly specified mechanism which identifies the conditions under which lobbying can be influential and how influence depends on the strategy chosen while the strategy choice depends on the expected influence.

The choice of lobbying strategies has received much attention in the literature. Two sets of variables are commonly highlighted in explanations of interest group strategies: Group characteristics, such as their type and resources and the issue context (Baumgartner and Leech 1998). Empirical studies which analyze interest group behavior often survey interest groups to elicit information about the determinants of their behavior (see Baumgartner and Leech 1998 for an extensive survey). Some studies elicit the network structure or try to explain network formation as part of lobbying strategies (Heinz et al. 1993; Laumann and Knoke 1987). Other studies enhance our knowledge about lobbying tactics and systems, e.g. in the European Union (Greenwood 2007; Dür and Mateo 2016), the United States (Baumgartner et al. 2009) or in a comparative perspective (Mahoney 2008). These studies built on an impressive collection of data and describe patterns and tactics. However, much of the explanation rests on a classical distinction of interest group types.

Influencing policy outcomes is the ultimate goal of an interest group. In consequence, it is necessary to analyze the interest groups' actions from *their* perspective. When trying to explain interest group behavior it is vital to understand how the political system and its channels of influence look from the group's perspective. This book is a first attempt to embrace this perspective.

---

[1]The book is based on my dissertation with the title *Political Uncertainty and Interest Group Communication Strategies* (Köhler 2014).

I distinguish interest group lobbying strategies based on a set of structurally equivalent communication acts. Two aspects of these classes of lobbying activities are important here. First, interest group activities are conceived as communicative acts. This approach acknowledges that all activity has the potential to transmit information from interest groups to decision-makers. Second, the distinction between public and private communication is crucial. The main difference between the classes of activities is the mode of the communication, i.e. whether it is observable or not. I show that the approach can be used to enhance our understanding of interest group communication strategies beyond the classical approach based on the group types or the insider/outsider distinction. I, therefore, offer a complementary explanation which captures much of the variation in interest group activities which is not covered by the more classical approaches.

Two interrelated decisions have to be analyzed. When do interest groups become active? and what do interest groups do once they mobilize? Interest groups have a choice between many communication strategies (including noncommunication). My approach allows to develop a coherent framework to study the two interrelated decisions, which are usually studied in relative isolation. It asks how the world looks like from an interest group's point of view to improve our understanding of their actions.

My explanation of interest group strategy choice explicitly connects group characteristics and context variables. Particular emphasis is put on the role of interest group preferences and the bargaining environment, in which lobbying is embedded. The approach has implications for how to best assess lobbying strategies. Identifying the causal mechanism which links interests and institutions is central for the understanding and explanation of interest group behavior. Based on a formal model, I demonstrate, that lobbying strategies can be understood and explained when situated in the context of legislative bargaining.

Interest groups seek to influence public policies. The question of why interest groups use a specific strategy to influence a policy is often conflated with the question why they use a specific tactic to influence a decision-maker. Political decisions are the result of political bargaining. Influencing a political actor is a means to influence an outcome and the activity has to be understood as such. This point is evident in the analysis of König (1992). His analysis rests on a series of Coleman models which are purely probabilistic in nature. While serving as a first approximation, the models lack a strategic component and the precision required to identify the causal mechanism.

Contrary to this approach, I draw on signaling models of lobbying in the tradition of models, such as Austen-Smith (1993), Potters and van Winden (1992), or Sloof (1998). At the core of these models is the assumption of an informational asymmetry between the interest group and the decision-maker. Interest groups possess better information about how policies translate into policy outcomes. Decision-makers face uncertainty about this nexus. I call this uncertainty *fundamental uncertainty* as it relates to fundamental processes in the social and physical world. The uncertainty of political decision-makers about the effects of policies is one of the driving factors which shapes the demand for lobbying (Truman 1971; Esterling 2004).

The decision-making process is shaped by institutions. Therefore, one has to analyze institutions and the way they determine incentive structures for decision-makers. A novel approach to the study of interest group strategies along these lines is proposed. I provide an explanation of interest group behavior as the result of incentives for strategic action created by the link between group activity and public policies as mediated by the political process. Interest groups interfere with the bargaining process of political actors. The institutions which structure legislative bargaining therefore determine how interest group activities translate into policies. This influence is necessarily incomplete, even if interest groups influence all political decision-makers. The reason is the conflict of interest of the decision-makers, which has strong implications for the willingness to lobby and the choice of strategy.

When choosing lobbying strategies, interest groups need to form an expectation about the likely policy outcomes. The final policy is agreed on in bargaining process between decision-makers, e.g. the chambers in a bicameral legislature or the coalition partners in a unicameral setting. Policies may vary because interest groups determine the distribution of information in the bargaining arena by choosing a specific lobbying strategy. Thus, interest groups themselves operate in an uncertain environment. They are unable to predict the exact outcome of the bargaining process when choosing a course of action. But they may be able to predict a range of possible or likely outcomes. I show both theoretically and empirically that this *process uncertainty* is an important determinant of interest group strategy choice.

My book starts with a discussion of the current state of our knowledge about interest group strategies and political bargaining, which serves to prepare the ground for the heart of the book, the formal model of interest group strategy choice. In particular, I demonstrate the importance of information and communication for lobbying. I then discuss the approaches to bargaining and how it creates uncertainty from an interest groups point of view.

The next step is the development of a formal model of interest group activity. The main novelty is the integration of a signaling model of lobbying with a model of veto bargaining. In contrast to most other formal models of lobbying, which were developed with the context of the American Congress in mind, the model explicitly focuses on a parliamentary system with high party discipline. In this model, interest groups can choose between sending a private message to one of the two decision-makers and sending a public message to both. The distinction is roughly equivalent to the classical inside and outside lobbying terminology. However, the implications are quite different.

Pro-change groups are more constrained in their lobbying behavior. The decision-maker which prefers a smaller move away from the Status Quo effectively determines the room for maneuver in the political system. This does not only affect the other decision-maker in terms of bargaining outcomes, but also the interest group's lobbying strategy. Pro-change groups are more likely to prefer private messages to the constraining actor over public messages, as they need to loosen the constraint. The more interest groups want to prevent a change of the Status Quo, the more likely they are to send a public message. They are unable to prevent the policy change, but they are able to reduce the variation of policies (i.e. the uncertainty), which makes them better off.

The model also demonstrates that under a wide range of circumstances, interest group influence on actual policy outcomes is very limited. However, they will reduce the variance of outcomes considerably. This raises the suspicion that many scholars are looking in the wrong place to find interest group influence by analyzing (expected) outcomes. Following Braumoeller (2006), looking for a causal effect on the variance of policies might be the right causal effect to look for.

I test the predictions of the model empirically, using data on lobbying in Germany. One particular aspect of the German political system allows me to identify the causal effect of the conflict of interest between the interest groups and the decision-maker, as well as the expected policy outcome. The German constitution exogenously determines whether the second chamber has a full veto or not. This is based on the content of the bill and cannot be influenced by the political decision-makers. The relevant decision-makers from an interest groups point of view are thus either the two chambers or the coalition partners in the government. This provides variation in the expected outcomes and the ideological distance to the expected outcome which I exploit to identify the causal effect.

The data were collected by a collaborative research project and lead to a couple of publications (König 1992; Pappi et al. 1995; Knoke et al. 1996) who used network approaches to describe and analyze the decision-making structures and processes in the policy domain of labor and social policy from 1982 to 1989. The data contain information on interest group activities regarding specific bills in the domain of labor and social policy. No other available dataset contains all the information necessary to identify the causal effect and empirically assess the causal mechanism. At the same time, the data allow to control for alternative explanations. I use a subset of the data to test the theory which has surprisingly not received much attention.

The main predictions of the model hold in the empirical analysis. Interest groups who are opposed to a change of the Status Quo are less likely to send private messages and are more likely to send public messages, independent of their group type. As the theory predicts, the distance to the expected policy outcome and the compromising actor are the key variables to explain interest group strategy choice. I demonstrate that the choice of strategies are interdependent, which is strong evidence in favor of the model.

The results are robust to the choice of statistical model and also hold for alternative operationalizations of the dependent variable. I demonstrate that in terms of the determinants of behavior, the specific activities which are aggregated in the variable for public messages are substitutes. The activities which make up the private messages can be classified as substitutes as well. The potential of the approach to restructure the thinking on interest group strategies based on the classes of structurally equivalent activities is therefore shown to be a promising venue for future research.

REFERENCES

Austen-Smith, D. (1993). Information and Influence: Lobbying for Agendas and Votes. *American Journal of Political Science, 37*(3), 799–833.

Baumgartner, F. R., Berry, J. M., Hojnacki, M., Kimball, D. C., & Lech, B. L. (2009). *Lobbying and Policy Change: Who Wins, Who Loses, and Why*. Chicago: University of Chicago Press.

Baumgartner, F. R., & Leech, B. L. (1998). *Basic Interests*. Princeton, NJ: Princeton University Press.

Braumoeller, B. F. (2006). Explaining Variance: Or Stuck in a Moment We Can't Get Out Of. *Political Analysis, 14*(1), 268–290.

Dür, A. & Mateo, G. (2016). *Insiders Versus Outsiders: Interest Group Politics in Multilevel Europe.* Oxford: Oxford University Press.

Esterling, K. M. (2004). *The Political Economy of Expertise: Information and Efficiency in American National Politics.* Ann Arbor: The University of Michigan Press.

Greenwood, J. (2007). *Interest Representation in the European Union* (2nd ed.). Houndmills, Basingstoke: Palgrave Macmillan.

Heinz, J. P., Laumann, E. O., Nelson, R. L., & Salisbury, R. H. (1993). *The Hollow Core.* Cambridge: Harvard University Press.

Knoke, D., Pappi, F. U., Broadbent, J., & Tsujinaka, Y. (1996). *Comparing Policy Networks.* Cambridge: Cambridge University Press.

Köhler, S. (2014). *Political Uncertainty and Interest Group Communication Strategies.* Dissertation, University of Mannheim.

König, T. (1992). *Entscheidungen im Politiknetzwerk.* Wiesbaden: Deutscher Universitäts Verlag.

Laumann, E. O., & Knoke, D. (1987). *The Organizational State: A Perspective on National Energy and Health Domains.* Madison: University of Wisconsin Press.

Mahoney, C. (2008). *Brussels Versus the Beltway: Advocacy in the United States and the European Union.* Washington, DC: Georgetown University Press.

Pappi, F. U., König, T., & Knoke, D. (1995). *Entscheidungsprozesse in der Arbeits- und Sozialpolitik.* Frankfurt/M.: Campus Verlag.

Potters, J., & van Winden, F. (1992). Lobbying and asymmetric information. *Public Choice, 74*(3), 269–292.

Sloof, R. (1998). *Game-Theoretic Models of the Political Influence of Interest Groups.* Boston: Kluwer.

Truman, D. B. (1971). *The Governmental Process. Political Interests and Public Opinion* (2nd ed.). New York: Knopf.

CHAPTER 2

Fundamental Uncertainty: The Demand for Information and Interest Group Activities

Abstract In this chapter, I discuss the approaches to understand interest groups strategies. I show that both theoretically and empirically one of the major factors for lobbying is the demand for information. It is created by the fact that decision-makers decide in uncertain situations. Empirical evidence demonstrates that the provision of information is one of the main activities of interest groups.

Keywords Demand for information · Uncertainty · Lobbying
Lobbying tactics

2.1   The Good, The Bad, or The Ugly?

Lobbying is an integral part of political processes. While it may be obvious in democracies, lobbying is also part of the political process in autocracies such as China (e.g. Steinberg and Shih 2012). Outside of academic circles—and sometimes even within—lobbying is seen as an evil. The media draw the picture of the lobbyist as a sketchy actor who extracts whatever he can from the political system. Seen in this way, lobbying is a zero-sum game that creates winners and losers. It is the mighty special interests, the story continues, who win at the expense of the public, which is defenseless and weak. The stylized representation is misleading in the extremity of its conclusions. It is true to the extent that all politics is about the distribution of values (Easton 1965a, b).

In academic analyses, lobbying is often seen more favorable or at least more neutral. I discuss the main approaches to lobbying in this chapter. It serves as the basis for the formal model which I develop in Chapter 4.

The first question to clarify in a scientific analysis of lobbying activities is what exactly is lobbying? Thomas (2004) highlights two components: (1) interaction: A decision-maker and an interest group interact. The interaction may be direct or indirect. (2) Goal orientation: The interest group's goal is to influence public policies, either today or in the future.

An important aspect in this definition is the group. The standard conception of lobbying is group based. An *interest group* is an institutionalized interest. As such it could be an association of individuals or organizations with one or more shared interests whose goal is to influence public policy to serve its interest (cf. Thomas 2004, 4). In a very broad sense, all citizens are members of an interest group on some issues as well as members of the public on all other issues (Lohmann 2003).

Interest groups are corporate actors that aggregate the interests of individuals. Who belongs to the group is defined on the basis of a common or shared interest. An example is the common interest of car producers regarding environmental standards for cars. To the extend that stricter environmental regulation creates costs for research and development and changes in production structures, all car producers have an interest in avoiding stricter regulation.

This was evident for example at the European level, where the European Association of Car Producers (ACEA) and the individual companies tried to water down the regulations during the negotiations about the Auto-Oil I programme which finally led to the Euro 4 norms for pollutant emissions of cars. The car makers also tried to gain a burden sharing agreement by extending the regulation to other sectors, such as the oil industry (hence the name Auto-Oil programme).

A commonly defined interest does not preclude conflicts of interest. The same group of car producers was rather divided on the question of the reduction of CO_2 emissions as the French and Italian producers are specialized in smaller cars. Small cars emit less CO_2 which gives them a comparative advantage in this field, compared to German car producers. However, the common interest is reasonably well defined and stable to distinguish car producers as a group. The same holds true for many other interests, for example, employers *vs.* workers who can be defined as an interest in the sense of the definition above.

Note also that while most people conceive of interest groups as actors related to the private sector, there is also governmental or public interest

Table 2.1 Theoretical and empirical approaches to lobbying

Theoretical approaches		Empirical approaches	
Formal models	Groups	Groups	Network
Influence function	Pluralism	Surveys	Description
Signaling	Corporatism	Case studies	Network formation
All-pay auctions	Olson	Comparison	
Menu auctions	Resource exchange		
	Population ecology		
	Punctuated equilibrium		

groups. Thomas (2004, 3) highlights the importance of interest groups which are representing government itself. This is particularly important for government agencies at the state and federal levels. The idea of public interest groups is consequential for my empirical treatment of interest group activities where I consider party organizations without direct legislative powers and ministries as interest groups.

The interest group needs to delegate the right to act on behalf of their interests to natural person(s). We call them lobbyists. They usually perform a variety of functions including contacts with officials, monitoring of politics, developing lobbying strategies and providing the manpower to run a lobbying campaign on behalf of the interest (Thomas 2004).

A lobbyist may be employed by the interest group (an in-house lobbyist) or working for a law-firm or consultancy which offers its services and contacts to paying customers (a 'hired gun'). While the latter are commonly found in Washington and Brussels, they are less prominent in Berlin.

One central activity of lobbyists when contacting officials is the attempt to influence legislative outcomes by means of (strategic) communication. In the following, I summarize the current theoretical and empirical knowledge on the relationship of interest groups and law-making in representative democracies. Table 2.1 summarizes the different approaches in the literature which will be discussed.

2.2   THE GROUP APPROACH TO LOBBYING

The conception of interest groups is rooted in a specific approach to politics. In this approach, society is seen as an aggregate of—possibly crosscutting—interests which are in competition for resources and public policies. The approach of the pluralist school based on Bentley (1908) and Truman (1971) anchors interest representation in democratic theory. The approaches usually follow a systemic or macro perspective, investigating how interest representation is institutionalized. Main concern is the effect that interest representation has on the democratic quality of political decisions.

Bentley (1908) was one of the pioneers in the study of interest representation. The basic argument of Bentley and other pluralists such as Truman (1971) is that the competition between societal groups that pursue their interest enhances democracy as it allows the articulation (input legitimacy) and thus the representation (output legitimacy) of these interests. Critical for the representation of interests is the level of access that governmental institutions grant (Truman 1971, Chapter 11).

The demand for information is a major driving force of lobbying in the pluralist school. It stems from what I call *fundamental uncertainty*, which is a result from imperfect knowledge of fundamental mechanisms and processes by which policies translate into outcomes in the physical world. Truman (1971) explicates the decision-makers' dependence on information (p. 331ff.) and claims that the problem has two sides. First, politicians require technical knowledge. Politicians also require political knowledge about of the relative weight of competing interests and how they will be affected by a policy.

The dependency is driven by politicians' desire to make a better decision in an uncertain world. This decision may be motivated by the desire to stay in office or by the desire to implement a specific policy for ideological motives. Both lead to the same dependency of information which is provided by the interest group which has an informational advantage.

It did not take long until criticism arouse. Olson (1965) argued that an equal representation of interests is not generally given. Problems of collective action hamper the articulation of diffuse interests such as consumers' interests. In consequence, these interests will not be represented at all. Concentrated interests such as producer's interests, however, are well-suited to overcome the problems of collective action. The result is a bias in the representation of interests and hence distorted public policies. Schattschneider (1975) similarly highlights the biases in mobilization patterns of interest groups although his approach of counting groups to assess bias has been the subject of criticism itself (Lowery and Gray 2004).

One of the reactions to the criticism was to drop the assumption that interests would balance each other. As a result, the view of the state shifted from being a neutral broker of interests to highlight the responsibility of enabling specific public interests to reduce biases. The approach of brokering interest groups to ensure a level playing field is known as neo-pluralism (Manley 1983; McFarland 2007). Neo-pluralist ideas find their expression in the monetary and institutional support for the European Trade Union Confederation (ETUC) and other public interest groups by the European Union (Greenwood 2007).

Becker (1983, 1985) went into a different direction. His approach is direct response to Olson's critique. Still inclined to a group conception of interest representation (Potters and Sloof 1996), he shows that groups will be able to overcome collective action problems if the biases of public policies become too large and the costs from maintaining these imbalances are higher than the costs of collective action. Becker's work was a major step toward an analytical (or positive) approach to interest representation. The model was adapted to a two-dimensional context by Ward (2004).

2.3   THE STRATEGIC PERSPECTIVE: FORMAL MODELS OF LOBBYING

The positive approach to interest representation started to emerge in the 1970s when the focus shifted toward the analysis of practices and impacts of interest groups on regulation (e.g. Stigler 1971; Peltzman 1976). These approaches were still in line with Olson's by arguing that the demand for regulation is driven by rent seeking groups such as producers. However, they focused on regulation and raised issues of asymmetric information and agency. The approach is based on a game theoretic framework and highlights utility maximizing politicians and interest groups and their *strategic* interaction with voters.

Afterward, the formalization of theorizing about interest groups gained momentum. The analytical approach is more interested in analyzing specific mechanisms of interest intermediation. This mechanism is modeled explicitly with game theoretical tools by decomposing the process into a strategic interaction of rational actors who are constrained by institutions. These actors are lobbyists and politicians. While the model analyzes a specific lobbying contact, it implicitly rests on the idea of group politics. While preferences are primitives of the models, i.e. they are assumed as given and stable, a justification for actors preferences in the real world has to make use of ideas developed in the group approach to politics.

The formal literature comprises several other types of models. First, there is the so-called *common agency* models (e.g. Bernheim and

Whinston 1986a, b; Grossman and Helpman 1994, 1996; Dixit et al. 1997). These models seek to explain interest group influenced by the benefits politicians can extract from interest groups. For maximizing these benefits, the decision-maker weighs interest groups according to their contribution (e.g. campaign spending, votes). The policy reflects the relative importance of interest groups *for the decision-maker*.

Closely related are models of vote buying and campaign contributions. Here, interest groups can win a prize (usually a policy or a favorable vote) by competitive bidding. The policy is effectively for sale, as the highest bidder wins at the expense of the other bidders (Hillman and Riley 1989). In some cases, bidding is seen as a means to buy access to a decision-maker. Austen-Smith (1997, 1998) shows that interest groups may use campaign contributions to buy access, but that influence rests on the provision of information. Also Cotton (2012, 2016) shows that bids may help in the competition for attention.

The efficacy of campaign contributions as a way of buying votes has been questioned early on. Langbein (1986) finds that informational lobbying of gun-control groups was quite successful, while there was negligible influence of campaign contributions. Hall and Wayman (1990) show that campaign contributions don't buy votes but may buy a higher level of involvement of the legislator. Wright (1996, 85f.) finds little empirical evidence for vote buying in the US Congress. The best predictor for a legislator's vote is his ideology. Wright concludes that information is the most important aspect of lobbying.

Milyo (2001) shows that incumbents care more about policies than about campaign contributions and will therefore be more interested in information. Campaign contributions in Germany are strictly regulated and large donors usually donate to all parties. It is therefore safe to ignore the donations part of lobbying. A strong indication for the legislature's dependency on information in Germany is the finding that the Bundestag receives more information than it sends (Knoke et al. 1996). A recent study by McKay (2012a) finds little influence of organizational resources on lobbying success. All these findings suggest that information plays a dominant role in lobbying.

The starting point of approaches to lobbying based on the information is the informational asymmetry between the uninformed decision-maker and the informed interest group. The asymmetry creates a principal-agent problem which is modeled in cheap talk and signaling models. I call this approach the informational approach to lobbying. The tools for analyzing the effects of asymmetric information were borrowed from the field of information economics and were originally developed for analyzing market interactions (Birchler and Bütler 2007; Hirshleifer and Riley 1991; Macho-Stadler and Pérez-Castrillo 2001).

The analytical tools turned out to be useful for analyzing the interaction of decision-makers and interest groups. While mainly based on the US system (van Winden 2008) the approach can be used in a fruitful way to understand lobbying in Germany and other parliamentary systems with strong party discipline, as I will demonstrate in this book.

Information has two main forms: knowledge and news. *Knowledge* refers to a *stock* of accumulated information, i.e. knowledge is information about aspects of the world based on the experience, data or other sources (e.g. interest groups). *News* denotes a *flow* of such information which adds to the stock of knowledge one possesses (Hirshleifer and Riley 1991, 167f.). It seems evident that absent any changes in the stock of information (knowledge) and absent any changes in the circumstances to which the information in stock is applicable, no change in behavior should be expected.

Decision-makers valued information because it helps to make better decisions. The value of information (i.e. of an interest group) can be defined as the increase in utility a decision-maker expects from optimally reacting to the information (the news) (Birchler and Bütler 2007, 32). It is important to note that information is valuable because it allows to get an outcome which would be infeasible without it. Seen in this way having more information can never be harmful as the decision-maker could simply ignore it (Birchler and Bütler 2007).

The informational approach to lobbying consists of several types of models which are closely related to each other. The standard models are *cheap talk* and *signaling* models. They both model information transmission from interest groups to decision-makers. The basic assumption is the existence of an informational asymmetry between the interest group and the decision-maker. The fact that interest group's interest is in general not fully aligned with the decision-maker's interest creates the possibility that the interest group misrepresents information. The models, therefore, seek to establish the circumstances under which the decision-maker can trust the information provided by the interest group, i.e. they seek to explain when lobbying can be influential.

Signaling models can be subdivided into three different approaches: *cheap talk and one shot signaling games, reputational games* (Sloof 1998;

Sloof and van Winden 2000), also Bernhagen (2008b), and *counteractive lobbying* (Austen-Smith and Wright 1992, 1994, 1996).

Models of *pressure politics* form a variation of informational models. The general logic is similar to a signaling model. Pressure is modeled as a transmission of information. The main difference is that—by transmitting information—interest groups can impose costs on policymakers (Sloof and van Winden 2000; Potters and van Winden 1990; Bernhagen and Bräuninger 2005). In these models, the structural power of interest groups is potentially stronger due to the ability to impose costs on decision-makers.

Access is a prerequisite for any direct attempt to influence a decision-maker.[1] Activities to gain access are thus instrumental for (and logically prior to) the sending of the information itself. In my analysis, I completely ignore activities carried out in order to gain access and treat access as given. As Austen-Smith (1998) has demonstrated, interest groups might buy access by means of campaign contributions but lobbying has to be understood as information transmission.

Interest groups try to influence policy *outcomes*. The key for a successful lobbying campaign is to influence the political actors which have the constitutional power to make binding decisions. In modern democracies, legislatures and governments are endowed with these powers. According to the terminology of Lasswell and Kaplan (1950), information is the base of influence which interest groups can use to influence politicians. Simon (1953) notes that influence and its bases are not the same. When trying to assess influence, both have to be studied separately.

Formal models that deal with the economics of information transmission model mechanisms of how the transmission of information leads to specific outcomes. They implicitly model a mechanism how interest groups' influence base (information) translates into influence (change in policies). Many models were explicitly created to analyze the informational aspects of interest representation. One important finding is that information understood as an influence base cannot always be used to influence outcomes.

The classical paper in the field is Crawford and Sobel (1982) who provide the general rationale in their model on cheap talk. One of the major findings of Crawford and Sobel is that the conflict of interest between the sender and the receiver determines the possibility of information revelation. In

---

[1] Note, however, that indirect attempts such as making public statements are possible without access.

particular, the closer the interests of the sender and receiver are aligned, the higher the likelihood for influential communication.

The common starting point of this literature is the idea of an informational asymmetry between an informed sender (the interest group) and an uninformed decision-maker. The sender is better informed about the state of the world, which is a random variable from the decision-makers point of view. After receiving a message, the decision-maker takes an action (e.g. implements a policy) which affects the welfare of both actors.

Influential communication in the sense of the model means that the decision-maker can update the beliefs about the state of the world and therefore chooses a better policy compared to an uninformed choice. Separating equilibria of the models are situations in which informative communication is possible.

Crawford and Sobel's standard model has been refined along several lines. The most important variations are in the number of players, the cost-structure of the message and the dimensionality of the policy space. Table 2.2 provides an overview over main features of the models. Austen-Smith and Banks (2000) analyze cheap talk and signaling models and demonstrate the close relationship between the models in the context of the spatial model.

Some approaches deal with more than one sender (Lohmann 1995; Grossman and Helpman 2001; Battaglini 2002). Lohmann (1995) analyzes how information is aggregated. She shows that if the information is a collective good, i.e. each interest group possesses a piece of information which only makes sense in the aggregate, information transmission is possible and the free-riding problem is of minor importance.

Grossman and Helpman (2001, Chapters 4, 5) analyze a signaling model with a variety of configurations and show that successful communication between interest groups and a decision-maker in settings with two interest groups depends on the bias of the two groups compared to the decision-maker. Informative and influential communication is possible in settings where the bias is going in the same direction, i.e. both prefer policies more (or less) extreme than the decision-maker. Informative communication is usually not possible in cases where both interest groups have opposing biases, i.e. the preferences of one group are more extreme while the other group prefers less extreme outcomes. To derive these results, Grossman and Helpman use a simplified version of a model by Krishna and Morgan (2001).

Table 2.2 A Taxonomy of informational models of lobbying

Article	No. Senders	No. Receivers	Signal costs	Communication	Dimensions
Crawford and Sobel (1982)	1	1	none	private	1
Farrell and Gibbons (1989)	1	2	none	private/public	2
Potters and van Winden (1992)	1	1	yes	private	1
Austen-Smith (1993)	1	2	yes	private	1
Sloof (1998)	1	1	yes	private	1
Austen-Smith and Banks (2000)	1	1	yes/no	private	1
Krishna and Morgan (2001)	2	1	yes	private/public	1
Grossman and Helpman (2001)	2	1	yes/no	private/public	1/2
Austen-Smith and Banks (2002)	1	1	yes/no	private	1
Battaglini (2002)	2	1	none	public	2
Crombez (2002)	1	2	none	private	1
Bernhagen and Bräuninger (2005)	1	1	yes	private	1

In contrast, Battaglini (2002) analyzes a model of cheap talk with two interest groups in a two-dimensional space. He shows that—despite the group competition—a revelation of information is generally possible if information is two-dimensional. His result is independent of both the bias of the interest groups and the size of the conflict of interest between the groups and the decision-maker.

The standard cheap talk or signaling models usually highlight the importance of preference alignment. Departing from this argument, Austen-Smith and Wright (1994) analyzes interest groups' incentives to target a legislator with opposing interests asking the question when an interest group will be able to persuade a policy maker to change his position on an issue. I will show, however, that interest group do not try to persuade the decision-maker but may have incentives to communicate information to a more distant actor in order to relax constraints in the bargaining environment. The latter case might falsely be interpreted as counteractive lobbying.

Signaling and *Cheap Talk* models are usually institution free in the sense that there is simply one decision-maker who takes a decision. The approaches usually focus on the interaction of lobbyists and legislators. The search for mechanisms comprises the question when, how and if interest groups can actually have an impact on legislators (Austen-Smith 1997; van Winden 2008). The models allow to derive hypotheses about mobilization patterns and communication strategies only with respect to one venue. Problematic is also that the model will do a good job in predicting contacts with individual legislators, but cannot explain lobbying in a wider context.

Most analytical models stop here and do not analyze how influencing a legislator (or collective actor such as a parliament) translates into influencing the outcome of collective decision-making. The modus operandi is to use a short-cut by assuming that the targeted legislator implements the policy unilaterally (van Winden 2008).

Austen-Smith (1993) and Crombez (2002) introduce more institutional structure into their models. Austen-Smith (1993) analyzes the possibilities of an interest group to influence the policy outcome in an intra-chamber game between the floor and a committee. He shows that lobbying at the agenda-setting stage is more influential than lobbying at the voting stage. Crombez (2002) expands this model and applies it to the bargaining between the Council and the European Parliament in the European Union. He allows more flexibility in the institutional structure by assuming that the targeted chamber moves first. In this setup, the interest group has some leverage over the outcome of the political process as well as over the sequencing.

Both models combine a model of information transmission and a bargaining model. However, interest groups perfectly anticipate the outcome of their lobbying activity. The feature of perfect anticipation of consequences is quite unrealistic when we talk about political processes, which are inherently uncertain due to the bargaining involved and random events such as changes in public opinion. I will develop a model which adds some realism on this dimension by explicitly modeling the uncertainty of policy outcomes from the interest group's perspective.

Farrell and Gibbons (1989), in contrast, model a situation where one sender sends a (costless) message to two decision-makers who take actions independently from each other. The sender can choose the mode of the message by either sending a private or public message. One of the major results is that depending on the conflict of interest between the decision-makers it does matter whether the message is communicated privately or publicly.

I take a similar approach in this book. However, I show that once there is more than one decision-maker whose decisions are not independent, the interest group can use its informational advantage to induce an information distribution between the chambers which de facto change the set of possible policy outcomes. The interest group can use those strategies to eliminate policies from the feasible set and change the expected outcome of the bargaining. Crucial is the interest group's ability to choose the target and the level of privacy of its message.

Esterling (2004) analyzes lobbying from the perspective of many groups competing for influence. His analysis deals with a collective action problem, which arises from the information being spread out across interest groups. His argument rests on a strong notion of informational dependency of decision-makers. He claims that due to the complexities of public policies, interest group activity is determined mostly by policy ideas and expertise.

Esterling's approach incorporates aspects of Lohmann (1995) and Becker (1983). However, Esterling assumes that actors make decisions in line with prospect theory, while all the other approaches build on expected utility theory. One of his main findings is that those groups who are in favor of a policy are risk averse, those against the policy are risk loving. He uses this argument to explain mobilization and signaling patterns. My model provides an alternative explanation as for why pro-policy groups should be less likely to act upon an issue.

The assumptions underlying the informational approach to lobbying are not unique. In the same way, in which the approach borrows some assumptions from the pluralist school, other approaches borrow assumptions from the informational models. Hall and Deardorff (2006, 74), for example, distinguish three types of information which interest group provide to legislators: (1) Constituency interests and opinions, (2) Expertise, and (3) Political 'intelligence' which follows a similar logic as informational models.

The authors acknowledge information as an important resource of lobbyists, but they make a point based on a resource exchange argument. Interest groups serve as an extension of the legislator in the sense that they provide information as a legislative subsidy. The legislator buys free time for working on other projects or engaging in otherwise more productive activities by outsourcing research and information gathering to interest groups who are better informed and resourceful.

The argument of Hall and Deardorff is less powerful in parliamentary systems, where legislation is usually drafted within the respective ministries, rather than by members of the legislature. A drawback of the argument is the problem of explaining inside and outside lobbying in a unified framework. Hall and Deardorff conceptualize outside lobbying as an attempt to change a legislators opinion. This is in contrast to my argument, which highlights information transmission and provides a unified treatment of what I call private and public communication.

2.4 EMPIRICAL EVIDENCE—REPRESENTATIVE DEMOCRACY AND INTEREST GROUP ACTIVITIES

The approaches to lobbying which I have discussed so far are mainly theoretical. Others follow an empirical strategy to understand interest group activities. Many of the approaches are inspired or informed by the theoretical arguments. I discuss the approaches based on the distinction between mobilization and action.

The question of mobilization patterns is important, although the literature dealing with interest group mobilization is relatively small. Leech et al. (2005), for example, analyze how government activity stimulates interest group activity. They show that government activity in a specific issue area induces interest group activity. The authors attribute this finding to the demand for lobbying created by legislative activities. While they do not explicitly attribute this to the uncertainty legislators face, uncertainty, and the resulting demand for information is an alternative explanation for the finding. Leech et al. are more interested in long-term developments and they do not explicitly derive a micro-based mechanism which may explain the results.

A related approach is taken by Dusso (2010). He builds on a population ecology approach to identify the reasons for differences in group activity across issue areas. While he can show that it is possible to predict the number of groups who are actively lobbying, the approach is unable to explain actual interest group behavior.

The approaches thus serve well to predict mobilization patterns in the long run and across issue areas, but they do not allow us to infer what the groups actually do. A big issue is that interest group preferences are ignored. Likewise, interest group activities are not systematically analyzed. However, it does make a difference, whether a group is just actively monitoring a legislative process or spending lots of resources on trying to change the outcome by lobbying.

Two important types of approaches can be identified in the empirical literature to explain interest group behavior. The first is the network approach. The second approach is empirical analyzes based on the surveys and observation of interest group activities.

2.4.1   Networks—Structure and Influence

Laumann and Knoke defined the standard for a network analysis of interest groups with the publication of their seminal book 'The Organizational State' (Laumann and Knoke 1987). They follow a structural approach to influence in the policy domains of health and energy in the USA. They ask how influence in specific policy events is intermediated in networks of resource and information exchange. Two aspects are crucial for actors to be considered influential. They need to have an interest and they need to be endowed with resources. The whole theoretical approach contains a feedback loop where policy outcomes affect the interest of actors in future events.

Interest groups were surveyed to elicit their behavior regarding important legislation in their respective domain of interest. However, interest group activities per se were not analyzed. Laumann and Knoke (1987) develop a statistical model of event participation. They explain this participation by network structures, i.e. an interest group's position in the network.

Problematic is that all actors are treated as equals. Their influence on events is essentially a function of their resources, interests, and reputation. But given the institutional power of legislatures and governments they are more consequential for policy outcomes than interest groups. Based on the idea of an exchange Laumann and Knoke (1987), Pappi et al. (1995) and Knoke et al. (1996) use Coleman models to analyze influence. These models are stochastic and lack a clearly specified causal mechanism. This is unfortunate as networks cannot produce outcomes themselves. When we argue that networks have an effect, this effect must be due to the ways in which networks influence how individuals act. This requires us to specify the mechanisms (Hedström 2008).

Only political decision-makers have veto power. Other groups lack the power to veto a decision. This point has been raised by König and Bräuninger (1998) and Bräuninger and König (2004). In both studies, the authors distinguish between political decision-makers and interest groups based on the observation that there is a power difference between the two types of actors. This allows to test alternative theories against each other in order to explain the formation of policy networks. One of the main explanatory variables is a measure of similarity which captures the alignment of actors preferences and is close to the ideas expressed in the formal models of strategic communication. The similarity of actors is a good predictor of network ties. Carpenter et al. (2004) analyze the incentives of interest group to share information in the US policy domain of energy and health politics. They also find that preference alignment matters.

Network approaches feature two of the most important factors for testing my theories as main ingredients. They identify actors and their interests. In addition, they observe their actions and analyze power relations between actors (Heinz et al. 1993). The latter are fixed in my model, political actors have the power to take binding decisions, interest groups have limited power over outcomes due to the possession of information.

Knoke et al. (1996) also discuss influence tactics. However, they only show that there are significant differences across the countries which they study. They do not distinguish the strategies or seek to explain them. I distinguish strategies according to a functionalist argument and aggregate them into equivalence classes. This allows me to add some structure to the analysis of interest group behavior which I can explain with the help of a theoretical model. Knoke et al. also ignore institutional aspects. The strength of my approach is that I explicitly account for institutional factors.

One of the most important explanatory factors in the studies of networks is the structure of issues and interests. This is the biggest difference to my approach because I argue that it is rather the institutional structure which defines how interests can be aggregated and how information can be translated into influence.

The models used by Pappi et al. (1995) seek to explain access to decision-makers and the control which interest groups are able to exercise over outcomes based on the trading their resources. They argue that agents' efficacy can be measured by their voting power. This, in turn, then drives the resource demand and supply concerning the issues. Their models create valuable insights about event participation and control but they ignore strategic incentives. It would not make sense for an actor to invest resources in an agent who is constrained in a way, that he is immaterial to the outcome, *independent* of his voting power. The argument of Pappi et al. (1995) is thus not necessarily incompatible with my arguments, but it does not go far enough in explaining actions.

2.4.2   *Lobbying Activities*

Many authors distinguish different types of lobbying tactics. Kollman (1998) for example, distinguishes *inside lobbying, outside lobbying* and *organizational maintenance*. Where *inside lobbying* refers to communication or interaction with decision-makers. *Outside lobbying* refers to public appeals to actors outside the policymaking community. *Organizational maintenance* refers to activities which help the group to overcome collective action problems to ensure its survival.

Gais and Walker (2001) analyze the strategies which interest groups choose to influence politics in the United States. They distinguish two broad categories: *inside vs. outside* lobbying. Gais and Walker claim that these strategies are relatively inflexible in the sense that groups choose strategies according to group characteristics and goals which do not change much. The strategy formation is analyzed from the viewpoint of interest groups maximizing the chances of survival. This creates an organizational logic where the choice of strategy is determined by the groups desire to survive. Influence and policy goals are of secondary concern. They matter only to the degree to which they affect the likelihood of group survival.

In this line of reasoning, Gais and Walker identify several important factors. The factors are (1) a conflicting environment, (2) characteristics of group membership, (3) organizational resources, and (4) the nature of financial support. These factors determine the choice of instruments early in the development of interest groups and are institutionally embedded in the group structure once the structure is fixed. The choice of lobbying tactics is genetically written into the structure of the group.

The basis for the argument is a survey of interest group activity. The groups were asked about the relative importance of several tactics. These were clustered into inside and outside strategies according to the type of activity. This picture, however, is necessarily incomplete, as it fails to explain the specific choice of lobbying activities in a given process.

It is, for example, at odds with the analysis of Hall and Reynolds (2012) who show that outside strategies often have inside ends. For example, media advertisements are targeted to specific legislators. This implies a *strategic* choice of interest group tactics. This strategic dimension is ignored by Gais and Walker (2001) who explain strategy choice based on purely organizational needs. While interest groups are seen as rational actors, the rationality is defined with respect to the overarching goal of survival.

Binderkrantz (2008) is another example. Her argument is based on the resource exchange framework. She argues that the group type, i.e. specific characteristics of the interest group determine strategy choice. This is a recurring theme, but again, once one takes a strategic perspective, this picture is necessarily incomplete.

Surveying interest groups to elicit information is a standard approach in empirical research on lobbying. Other famous examples in this line of research is the work of Schlozman and Tierney (1986) who highlight the role of organizations' resources for the choice of strategy. According to their study, it is the identity and characteristics of the group which drive tactic choice.

Schlozman and Tierney (1986) argue that the exact type of public action may depend on the resources at the disposal of the group. They give various examples of how different organizations can use strategies that are in correspondence with their budget. On the contrary, I argue that the relevant distinction is not so much whether groups organize a demonstration or buy airplay on radio and television stations. It is whether the message is publicly observable or not. Moreover, for the distinction of whether strategies are effective, only relative costs matter. The politician can deduce information from a groups effort and mobilization of resources. Ceteris paribus, rich interest groups have to invest more in order to be convincing. The material wealth is therefore not decisive.

Schlozman and Tierney (1986) argue further that once the arena defined, the choice of lobbying tactics is automatically narrowed down. This is true, but often the choice of tactic within a specific venue is embedded into a larger lobbying strategy designed to influence politics *across* lobbying venues. This strategic calculus becomes even more important, the more interest groups face scarce resources and time constraints to lobbying. It will also depend on the relative importance of the venue in the decision-making process as I will show in my model.

In a unicameral system, lobbying the parliament is different from lobbying one of the chambers in a bicameral system. It is problematic to explain the choice of lobbying tactics within a venue in isolation from the decision-making processes involving several venues. It is vital to embed the explanation of interest group tactics into a wider strategic analysis. I provide a new approach to this problem using a formal model as a starting point. In all these studies, a main point which is overlooked by the author(s) is the question of a structural equivalence of tactics.

Many of the abovementioned authors show that interest groups choose a mixture of tactics (Schlozman and Tierney 1986; Kollman 1998). What is still not well understood is the choice of lobbying strategy, i.e. why do interest groups choose a specific mixture of lobbying tactics. One common misunderstanding which hampers progress in the analysis of interest group strategies is the claim that that access and influence are inseparable (e.g. Schlozman and Tierney 1986).

Access is neither necessary nor sufficient for being influential. This point has been highlighted many times in variations of the informational model of lobbying (e.g. Grossman and Helpman 2001). In addition, Schlozman and Tierney (1986) stress the importance of equal access, i.e. the policymakers need to hear both sides of the story. This is somewhat surprising, given the many biases in the communication situation which may affect the likelihood of convincing the decision-maker. In addition, Krishna and Morgan (2001) have shown theoretically that the decision-maker may not be able to draw inferences on the effects of policies if he meets groups with opposing biases. Empirically, the evidence is also thin, as there is no evidence that indirect strategies are a fall-back option for groups who lack direct access to political decision-makers (Binderkrantz 2005).

There is empirical evidence for the fact that preference alignment between legislators and interest groups matters outside the realm of network analyzes. Hojnacki and Kimball (1998), for example, analyze which legislators in a committee of the House of Representatives are targeted by interest groups. One of their major findings is, that preference alignment matters. The closer the preferences, the more likely the legislator is targeted by an interest group. Other factors they highlight are available resources and the issue context. Lobbyists are aware of the special positions of some legislators. For example, they are more likely to lobby committee chairs. The results suggest a fair amount of strategic behavior of interest groups which is not reflected in the theoretical argument in many studies.

In a related paper, Hojnacki and Kimball (1999) analyze the choice of direct *vs.* indirect lobbying tactics. They argue that direct tactics are more likely to be used in committee as indirect tactics are serving the purpose of group maintenance, an argument similar to Gais and Walker (2001). They also find that interest groups target their friends in committee. The analysis builds on a dyadic data structure which is closely related to network approaches. In this sense, both papers are in a line of heritage of the network approach.

Kollman (1997) raises the issue of friendly and unfriendly (confrontational) lobbying. He claims that the high preference alignment between committee members and interest groups is an effect of self-selection. The committee members and the interest groups have similar biases to begin with. This bias is caused by the issue area. Liberals are more likely to serve in a committee for social policies. Therefore, the committee will be biased relative to the floor. The same bias will be present in the organized groups in the respective area.

Another explanation for the preference alignment phenomenon has been provided by Austen-Smith and Wright (1996) who claim that the observed preference alignment is due to counteractive lobbying. That is, measurement takes place after the interest group has changed the mind of the legislator and, therefore, we find preference alignment. Their argument is based on a formal model by Austen-Smith and Wright (1994).

Victor (2007) uses a strategic approach to explain interest group activities. She argues that the legislative context matters and that interest groups respond to differences in incentives which are created by different circumstances. One constraint is the salience of the issue. Victor measures salience as whether or not the issue is important for the public. However, unless one assumes that interest groups want to send messages to their constituency, public awareness should matter less than group-specific salience. If groups do not care about the issue, they will not lobby even if public awareness is high. This fact has already been highlighted by Knoke et al. (1996).

Victor also argues that lobbyists are less constrained when a bill is referred to several committees—compared to a situation where a bill is referred only to one committee—as the number of possible access points increases. But obviously, not all committees are created equal. Usually, there is one leading committee which makes the final decision whether to submit the bill to the floor or shelve it. Some committees may only perform an advisory function. It is therefore not clear that truly strategic interest groups are less constrained by the involvement of more committees. There may be a simple demand effect. The two facts are observationally equivalent in empirical studies.

What is lacking in all these approaches is a clear understanding of what interest groups actually do, as the context is *not* independent from the group's activity. The strategic perspective is key to this understanding. There is a strong need for theoretical models, which make predictions about group behavior that are testable and rule out alternative explanations to the extent possible.

McKay (2012b) argues that what she calls 'negative lobbying' matters for the success of a bill. Negative lobbying is lobbying against a bill. She shows that stronger opposition decreases the likelihood of a bill to be adopted. In my theoretical model, I will show that what she captures in her analysis is not about 'negative' lobbying per se. Her observations are a direct consequence of the constraints the decision-making context imposes on actors. Credible and meaningful communication is more likely for interest groups which oppose a change of the Status Quo.

Another type of argument is made by Michalowitz (2007). She analyzes interest group influence in the European Union and distinguishes two types of influence. She defines 'directional influence' as an attempt to change the core of a law, while she defines 'technical influence' as influence which is constrained to changing details of a bill without touching the core. Technical influence rests on technical expertise. She demonstrates by means of three case studies, that in the European Union interest groups can mainly exert influence of the second type.

This is a potential difference to the United States, where 'directional influence' seems easier to achieve. In general, all parliamentary system can be expected to be more prone to allow for 'technical influence' as governments usually control the majority in parliament. Once the government is dedicated to a bill it usually becomes law.

So far I have mainly discussed the demand for lobbying and inside strategies. Schlozman and Tierney (1986, Chapter 8) describe the reasons for interest groups to take public action. They distinguish two types of goals to go public: *Persuasion* and *mobilization*. The first activity aims to convince an audience of the group's position, the second one serves to increase the pressure on politicians to adopt a particular policy. Even public strategies may have their specific targets. While Schlozman and Tierney (1986) highlight public opinion leaders, Hall and Reynolds (2012) analyze media outlets specifically targeting the districts of individual legislators.

Closely related but of course confined to specific types of interest groups are mobilization tactics. Goldstein (1999) claims that a major motivation for mobilization activities is to convey information about the electoral consequences which result from a politician's actions in congress. The argument rests on the assumption that politicians are office-seekers. As a result, they are susceptible to the preferences of voters regarding specific policies.

In addition to policy objectives, groups may have other objectives, for example their effort may be electorally motivated. However, policy objectives are most common motivational source.[2] The policy motive is short term in nature. The group only tries to influence a specific policy, not the long-term chances of changing policies by strengthening electoral prospects of candidates with similar preferences. Particularly, interesting is the finding that the committee level is important for mobilization activities (Goldstein 1999, 59). Following Goldstein (1999) and Kollman (1998) I will understand outside lobbying, i.e. mobilization and media communication strategies as means to influence the positions of legislators and thereby the outcomes of their actions. The mechanism which I highlight is nevertheless different.

All these approaches are venue unspecific in the sense that they either ignore the targets of lobbying activities or are restricted to one venue, e.g. the Bundestag. Holyoke (2003) highlights the point that venue specific analyzes of interest group behavior might be misleading as interest groups very often have the choice between several venues and lobbying intensities—a point which I also emphasize in the theoretical model.

Holyoke, for example, shows that interest groups take the strategies of their opponents into account and choose venues where they expect less counteractive forces. The argument of Holyoke (2003) stresses the importance of a strategic perspective on the behavior of other interest groups. In the model, the strategic incentives created by interacting with different decision-makers will be highlighted. Strategic behavior caused by interest

[2]See Goldstein (1999, Table 4.1 on p. 55).

group competition is not explicitly taken into consideration. What matters for the choice of lobbying activity in my model is the expected impact which lobbying a specific decision-maker will have on the outcome of the political process. The interest group's choice of strategy is thus driven by the targets likely role in the political process and the costs involved in lobbying.

Other studies deal with the question of venue-specific activities, often in a nonstrategic perspective. One of the earliest is Lindberg (1963) who shows that second chambers that represent more partisan interests (e.g. state interests) are targeted more often by lobbyists trying to prevent an outcome. While his argument rests on structural differences between the chambers, I argue that it is indeed the distribution of preferences between the chambers that matters. To the degree that the preferences of chambers are determined by the composition there will be a strong correlation between these two aspects.

The institutional determinants of venue choice are also analyzed by Naoi and Krauss (2009). Their major interest is to explain why some interest groups lobby legislators and why some lobby bureaucrats. A major determinant in their explanation is the degree of centralization of the political system. Their argument further states that the effect of centralization is contingent on the electoral system.

Binderkrantz (2003) analyzes how interest groups adapt their strategy in response to changes in the relative importance of parliaments. She shows for the Danish case that a more important role of parliament leads to more activity by interest groups in the parliamentary arena. At the same time, she does not observe a decline in influence attempts in other arenas, like the bureaucracy. A cornerstone of her analysis is the definition of parliamentary activities. She uses a wide range of parliamentary activities like controlling the government, parliamentary scrutiny, and public communication, as opposed to a narrow definition of parliament as a law-making body. Her definition of a lobbying strategy is quite close to the idea of venue shopping as coined by Baumgartner and Jones (1993). More importantly, she demonstrates that institutional powers matter for the choice of lobbying target.

Another approach is to analyze lobbying strategies based on the distinction between access and pressure politics (Binderkrantz 2005). Binderkrantz defines the former quite generally as access to decision-makers (which may be institutional actors). She defines pressure politics as the use of mobilization or media strategies. She creates an index for the different types of strategies and runs multivariate models to analyze strategy choice. One of her main findings is that outside strategies are not a fall-back option for interest groups who do not have access to decision-makers. This raises the question of the relationship between the two types of strategies which remains vague. One of the problems in this respect is the distinction between pressure politics and access. The distinction obscures the informational content of lobbying, which is an important functional aspect in the constitution of the lobbying strategy.

Other studies highlight the importance of group type. Binderkrantz (2008) classifies interest groups into three broad categories: Groups with *corporatist* resources, *public* interest groups and a reference category of *other* groups. The data collection is survey based and rests on a comparison of population averages of specific strategy uses. In a similar study, Dür and Mateo (2013) analyze the choice of interest group tactics in a comparative design including Austria, Germany, Ireland, Latvia, and Spain. Their main categories used to distinguish interest groups tactics are *inside vs. outside* lobbying. Dür and Mateo analyze the relative use of several activities which they aggregate into the two indices which represent a measure for interest group activities. They furthermore distinguish three types of interest groups: *Business, Citizen,* and *Professional* groups. They find that the group type is one of the main explanatory factors. However, they also argue that the effect of group type is conditional on resources and the issue context. Most noteworthy is their finding that the differences between strategy choices are smallest in mainly regulatory policy areas. Regulatory policies are the areas where legislators are the most dependent on expertize of interest groups (Michalowitz 2007).

There is also a range of studies on interest representation in the European Union (EU). Due to its many similarities to a bicameral political system the literature is potentially relevant. The empirical studies are usually conducted as case studies or studies based on the surveying interest groups, politicians and/or bureaucrats (Coen 2007). Either the case studies focus on a specific policy field, such as trade-policy (Dür 2008; Dür and de Bièvre 2007; Beyers and Kerremans 2007), climate policy (Gullberg 2008a, b), Public Health Policies (Princen 2007), or European integration (Leblond 2008) or they compare different political systems (Mahoney 2007a, b; Mahoney and Baumgartner 2008; Baumgartner 2007; Lowery et al. 2008). Beyers (2004) deals with the activities of interest associations in general.

Coen (1997, 1998), Eising (2004, 2007a) and Bernhagen (2008a) deal with the representation of Business interests in the European Union. In contrast, Smith (2008) tries to assess the circumstances under which environmental interest groups are influential. All studies find that gaining access is an important determinant of interest group activity. I would again highlight the fact that access is no guarantee to be influential. Given the nature and heterogeneity of the political actors in the EU, it may also be more important to establish ties in the first place. Thus, gaining access may be a higher priority goal.

Theoretical approaches to interest representation in the European Union often apply a market logic to the interaction of lobbyists and EU institutions to explain access. These approaches highlight the exchange of resources and especially the resource dependencies of the European institutions. Bouwen (2002, 2004a, b) applies a Trumanian logic of access to the European system. He argues that different EU institutions grant access to different interest groups (firms, national associations, and European associations) because they are dependent on different resources (usually information) that interest groups provide. In this logic, interest groups are the supply side of the market. Mahoney (2004) turns the logic upside down. She shows that governmental activity is crucial for the mobilization of interest groups in the European Union.

Eising (2007b) analyzes the supply side of this interaction. He analyzes why different interest groups target different EU institutions and what factors determine differences in lobbying strategies between different groups. His findings suggest that resource dependencies, institutional opportunities, and organizational capacities matter. Lobbyists seek to maximize their expected impact given the institutional constraints. I will develop the argument further by analyzing the institutional constraints which interest groups face and how they affect their strategies.

References

Austen-Smith, D. (1993). Information and Influence: Lobbying for Agendas and Votes. *American Journal of Political Science*, *37*(3), 799–833.

Austen-Smith, D. (1997). Interest Groups: Money, Information and Influence. In D. C. Mueller (Ed.), *Perspectives on Public Choice* (pp. 296–322). Cambridge: Cambridge University Press.

Austen-Smith, D. (1998). Allocating Access for Information and Contributions. *Journal of Law Economics & Organization*, *14*(2), 277–303.

Austen-Smith, D., & Banks, J. S. (2000). Cheap Talk and Burned Money. *Journal of Economic Theory*, *91*(1), 1–16.

Austen-Smith, D., & Banks, J. S. (2002). Costly Signaling and Cheap Talk in Models of Political Influence. *European Journal of Political Economy, 18*(2), 263–280.

Austen-Smith, D., & Wright, J. R. (1992). Competitive Lobbying for a Legislator's Vote. *Social Choice and Welfare, 9*(3), 229–257.

Austen-Smith, D., & Wright, J. R. (1994). Counteractive Lobbying. *American Journal of Political Science, 38*(1), 25–44.

Austen-Smith, D., & Wright, J. R. (1996). Theory and Evidence for Counteractive Lobbying. *American Journal of Political Science, 40*(2), 543–564.

Battaglini, M. (2002). Multiple Referrals and Multidimensional Cheap Talk. *Econometrica, 70*(4), 1379–1401.

Baumgartner, F. R. (2007). EU Lobbying: A View from the US. *Journal of European Public Policy, 14*(3), 482–488.

Baumgartner, F. R., & Jones, B. D. (1993). *Agendas and Instability in American Politics*. Chicago: University of Chicago Press.

Becker, G. S. (1983). A Theory of Competition Among Pressure Groups for Political Influence. *The Quarterly Journal of Economics, 98*(3), 371–400.

Becker, G. S. (1985). Public Policies, Pressure Groups, and Dead Weight Costs. *Journal of Public Economics, 28*(3), 329–347.

Bentley, A. F. (1908). *The Process of Government*. Chicago: University of Chicago Press.

Bernhagen, P. (2008a). Business and International Environmental Agreements: Domestic Sources of Participation and Compliance by Advanced Industrialized Democracies. *Global Environmental Politics, 8*(1), 78–110.

Bernhagen, P. (2008b). *The Political Power of Business: Structure and Information in Public Policymaking*. London: Chapman & Hall.

Bernhagen, P., & Bräuninger, T. (2005). Structural Power and Public Policy: A Signaling Model of Business Lobbying in Democratic Capitalism. *Political Sudies, 53*(1), 43–64.

Bernheim, B. D., & Whinston, M. D. (1986a). Common Agency. *Econometrica, 54*(4), 923–942.

Bernheim, B. D., & Whinston, M. D. (1986b). Menu Auctions, Resource-Allocation, and Economic Influence. *Quarterly Journal of Economics, 101*(1), 1–31.

Beyers, J. (2004). Voice and Access: Political Practices of European Interest Associations. *European Union Politics, 5*(2), 211–240.

Beyers, J., & Kerremans, B. (2007). The Press Coverage of Trade Issues: A Comparative Analysis of Public Agenda-Setting and Trade Politics. *Journal of European Public Policy, 14*(2), 269–292.

Binderkrantz, A. (2003). Strategies of Influence: How Interest Organizations React to Changes in Parliamentary Influence and Activity. *Scandinavian Political Studies, 26*(4), 287–306.

Binderkrantz, A. (2005). Interest Group Strategies: Navigating Between Privileged Access and Politics of Pressure. *Political Studies, 53*, 694–715.

Binderkrantz, A. (2008). Different Groups, Different Strategies: How Interest Groups Pursue Their Political Ambitions. *Scandinavian Political Studies, 31*(2), 173–200.

Birchler, U., & Bütler, M. (2007). *Information Economics*. London: Routledge.

Bouwen, P. (2002). Corporate Lobbying in the European Union: The Logic of Access. *Journal of European Public Policy, 9*(3), 365–390.

Bouwen, P. (2004a). Exchanging Access Goods for Access: A Comparative Study of Business Lobbying in the European Union institutions. *European Journal of Political Research, 43*(3), 337–369.

Bouwen, P. (2004b). The Logic of Access to the European Parliament: Business Lobbying in the Committee on Economic and Monetary Affairs. *Journal of Common Market Studies, 42*(3), 473–496.

Bräuninger, T., & König, T. (2004). Senden und Empfangen: Informationstransfer in Politiknetzwerken als Vermittlung von Verhandlungsvorschlägen. In C. H. Henning & C. Melbeck (Eds.), *Interdisziplinäre Sozialforschung. Theorie und empirische Anwendungen* (pp. 205–224). Frankfurt: Campus Verlag.

Carpenter, D. P., Esterling, K. M., & Lazer, D. M. J. (2004). Friends, Brokers, and Transitivity: Who Informs Whom in Washington Politics? *The Journal of Politics, 66*(1), 224–246.

Coen, D. (1997). The Evolution of the Large Firm as a Political Actor in the European Union. *Journal of European Public Policy, 4*(1), 91–108.

Coen, D. (1998). The European Business Interest and the Nation State: Large-Firm Lobbying in the European Union and Member States. *Journal of Public Policy, 18*(1), 75–100.

Coen, D. (2007). Empirical and Theoretical Studies in EU Lobbying. *Journal of European Public Policy, 14*(3), 333–345.

Cotton, C. (2012). Pay-to-Play Politics: Informational Lobbying and Contribution Limits When Money Buys Access. *Journal of Public Economics, 96*(3/4), 369–386.

Cotton, C. (2016). Competing for Attention: Lobbying Time-Constrained Politicians. *Journal of Public Economic Theory, 18*(4), 642–665.

Crawford, V. P., & Sobel, J. (1982). Strategic Information Transmission. *Econometrica, 50*(6), 1431–1451.

Crombez, C. (2002). Information, Lobbying and the Legislative Process in the European Union. *European Union Politics, 3*(1), 7–32.

Dixit, A., Grossman, G. M., & Helpman, E. (1997). Common Agency and Coordination: General Theory and Application to Government Policy Making. *The Journal of Political Economy, 105*(4), 752–769.

Dür, A. (2008). Bringing Economic Interests Back into the Study of EU Trade Policy-Making. *British Journal of Politics & International Relations, 10*(1), 27–45.

Dür, A., & de Bièvre, D. (2007). Inclusion Without Influence? NGOs in European Trade Policy. *Journal of Public Policy, 27*(01), 79–101.

Dür, A., & Mateo, G. (2013). Gaining Access or Going Public? Interest Group Strategies in Five European Countries. *European Journal of Political Research, 52*(5), 660–686.

Dusso, A. (2010). Legislation, Political Context, and Interest Group Behavior. *Political Research Quarterly, 63*(1), 55–67.

Easton, D. (1965a). *Framework for Political Analysis.* Upper Saddle River: Prentice Hall.

Easton, D. (1965b). *A Systems Analysis of Political Life.* New York: Wiley.

Eising, R. (2004). Multilevel Governance and Business Interests in the European Union. *Governance, 17*(2), 211–245.

Eising, R. (2007a). The Access of Business Interests to EU Institutions: Towards Élite Pluralism? *Journal of European Public Policy, 14*(3), 384–403.

Eising, R. (2007b). Institutional Context, Organizational Resources and Strategic Choices: Explaining Interest Group Access in the European Union. *European Union Politics, 8*(3), 329–362.

Esterling, K. M. (2004). *The Political Economy of Expertise: Information and Efficiency in American National Politics.* Ann Arbor: The University of Michigan Press.

Farrell, J., & Gibbons, R. (1989). Cheap Talk with Two Audiences. *The American Economic Review, 79*(5), 1214–1223.

Gais, T. L., & Walker, J. L. (2001). Pathways to Influence in American Politics. In J. L. Walker (Ed.), *Mobilizing Interest Groups in America*, Chapter 6 (pp. 103–121). Ann Arbor: University of Michigan Press.

Goldstein, K. M. (1999). *Interest Groups, Lobbying, and Participation in America.* Cambridge: Cambridge University Press.

Greenwood, J. (2007). *Interest Representation in the European Union* (2nd ed.). Houndmills, Basingstoke: Palgrave Macmillan.

Grossman, G. M., & Helpman, E. (1994). Protection for Sale. *American Economic Review, 84*(4), 833–850.

Grossman, G. M., & Helpman, E. (1996). Electoral Competition and Special Interest Politics. *The Review of Economic Studies, 63*(2), 265–286.

Grossman, G. M., & Helpman, E. (2001). *Special Interest Politics.* Cambridge, MA: The MIT Press.

Gullberg, A. T. (2008a). Lobbying Friend and Foes in Climate Policy: The Case of Business and Environmental Interest Groups in the European Union. *Energy Policy, 36*(8), 2964–2972.

Gullberg, A. T. (2008b). Rational Lobbying and EU Climate Policy. *International Environmental Agreements: Politics, Law and Economics, 8*(2), 161–178.

Hall, R. L., & Deardorff, A. V. (2006). Lobbying as Legislative Subsidy. *American Political Science Review, 100*(1), 69–84.

Hall, R. L., & Reynolds, M. E. (2012). Targeted Issue Advertising and Legislative Strategy: The Inside Ends of Outside Lobbying. *American Journal of Political Science, 74*(3), 888–902.

Hall, R. L., & Wayman, F. W. (1990). Buying Time: Moneyed Interests and the Mobilization of Bias in Congressional Committees. *American Political Science Review, 84*(3), 797–820.

Hedström, P. (2008). Studying Mechanisms to Strengthen Causal Inferences in Quantitative Research. In J. Box-Steffensmeier, H. E. Brady, & D. Collier (Eds.), *Oxford Handbook of Political Methodology* (pp. 319–335). Oxford: Oxford University Press.

Heinz, J. P., Laumann, E. O., Nelson, R. L., & Salisbury, R. H. (1993). *The Hollow Core*. Cambridge: Harvard University Press.

Hillman, A. L., & Riley, J. G. (1989). Politically Contestable Rents and Transfers. *Economics and Politics, 1*(1), 17–39.

Hirshleifer, J., & Riley, J. G. (1991). *The Analytics of Uncertainty and Information*. Cambridge: Cambridge University Press.

Hojnacki, M., & Kimball, D. C. (1998). Organized Interests and the Decision of Whom to Lobby in Congress. *The American Political Science Review, 92*(4), 775–790.

Hojnacki, M., & Kimball, D. C. (1999). The Who and How of Organizations' Lobbying Strategies in Committee. *The Journal of Politics, 61*(4), 999–1024.

Holyoke, T. T. (2003). Choosing Battlegrounds: Interest Group Lobbying Across Multiple Venues. *Political Research Quarterly, 56*(3), 325–336.

Knoke, D., Pappi, F. U., Broadbent, J., & Tsujinaka, Y. (1996). *Comparing Policy Networks*. Cambridge: Cambridge University Press.

Kollman, K. (1997). Inviting Friends to Lobby: Interest Groups, Ideological Bias, and Congressional Committees. *American Journal of Political Science, 41*(2), 519–544.

Kollman, K. (1998). *Outside lobbying*. Princeton, NJ: Princeton University Press.

König, T., & Bräuninger, T. (1998). The Formation of Policy Networks. *Journal of Theoretical Politics, 10*(4), 445–471.

Krishna, V., & Morgan, J. (2001). A Model of Expertise. *The Quarterly Journal of Economics, 116*(2), 747–775.

Langbein, L. I. (1986). Money and Access: Some Empirical Evidence. *The Journal of Politics, 48*(4), 1052–1062.

Lasswell, H. D., & Kaplan, A. (1950). *Power and Society: A Framework for Political Inquiry*. New Haven: Yale University Press.

Laumann, E. O., & Knoke, D. (1987). *The Organizational State: A Perspective on National Energy and Health Domains*. Madison: University of Wisconsin Press.

Leblond, P. (2008). The Fog of Integration: Reassessing the Role of Economic Interests in European Integration. *British Journal of Politics & International Relations, 10*(1), 9–26.

Leech, B. L., Baumgartner, F. R., Pira, T. M. L., & Semanko, N. A. (2005). Drawing Lobbyists to Washington: Government Activity and the Demand for Advocacy. *Political Research Quarterly, 58*(1), 19–30.

Lindberg, L. N. (1963). *The Political Dynamics of European Economic Integration*. Stanford: Stanford University Press.

Lohmann, S. (1995). Information, Access, and Contributions: A Signalling Model of Lobbying. *Public Choice, 85*(3–4), 267–284.

Lohmann, S. (2003). Representative Government and Special Interest Politics (We Have Met the Enemy and He Is Us). *Journal of Theoretical Politics, 15*(3), 299–319.

Lowery, D., & Gray, V. (2004). Bias in the Heavenly Chorus Interests in Society and Before Government. *Journal of Theoretical Politics, 16*(1), 5–29.

Lowery, D., Poppelaars, C., & Berkhout, J. (2008). The European Union Interest System in Comparative Perspective: A Bridge Too Far? *West European Politics, 31*(6), 1231.

Macho-Stadler, I., & Pérez-Castrillo, J. (2001). *An Introduction to the Economics of Information: Incentives and Contracts* (2nd ed.). Oxford: Oxford University Press.

Mahoney, C. (2004). The Power of Institutions: State and Interest Group Activity in the European Union. *European Union Politics, 5*(4), 441–466.

Mahoney, C. (2007a). Lobbying Success in the United States and the European Union. *Journal of Public Policy, 27*(1), 35–56.

Mahoney, C. (2007b). Networking vs. Allying: The Decision of Interest Groups to Ioin Coalitions in the US and the EU. *Journal of European Public Policy, 14*(3), 366–383.

Mahoney, C., & Baumgartner, F. (2008). Converging Perspectives on Interest Group Research in Europe and America. *West European Politics, 31*(6), 1253.

Manley, J. F. (1983). Neo-pluralism: A Class Analysis of Pluralism I and Pluralism II. *The American Political Science Review, 77*(2), 368–383.

McFarland, A. S. (2007). Neopluralism. *Annual Review of Political Science, 10*(1), 45–66.

McKay, A. (2012a). Buying Policy? The Effects of Lobbyists' Resources on Their Policy Success. *Political Research Quarterly, 65*(4), 908–923.

McKay, A. (2012b). Negative Lobbying and Policy Outcomes. *American Politics Research, 40*(1), 116–146.

Michalowitz, I. (2007). What Determines Influence? Assessing Conditions for Decision-Making Influence of Interest Groups in the EU. *Journal of European Public Policy*, *14*(1), 132–151.

Milyo, J. (2001). What Do Candidates Maximize (and Why Should Anyone Care)? *Public Choice*, *109*(1/2), 119–139.

Naoi, M., & Krauss, E. (2009). Who Lobbies Whom? Special Interest Politics Under Alternative Electoral Systems. *American Journal of Political Science*, *53*(4), 874–892.

Olson, M. (1965). *The Logic of Collective Action: Public Goods and the Theory of Groups*. Cambridge, MA: Harvard University Press.

Pappi, F. U., König, T., & Knoke, D. (1995). *Entscheidungsprozesse in der Arbeits- und Sozialpolitik*. Frankfurt/M.: Campus Verlag.

Peltzman, S. (1976). Toward a More General Theory of Regulation. *Journal of Law and Economics*, *19*(2), 211–240.

Potters, J., & Sloof, R. (1996). Interest Groups: A Survey of Empirical Models That Try to Assess Their Influence. *European Journal of Political Economy*, *12*, 403–442.

Potters, J., & van Winden, F. (1990). Modelling Political Pressure as Transmission of Information. *European Journal of Political Economy*, *6*(1), 61–88.

Potters, J., & van Winden, F. (1992). Lobbying and Asymmetric Information. *Public Choice*, *74*(3), 269–292.

Princen, S. (2007). Advocacy Coalitions and the Internationalization of Public Health Policies. *Journal of Public Policy*, *27*(1), 13–33.

Schattschneider, E. E. (1975). *The Semisovereign People: A Realist's View of Democracy in America*. Hinsdale, Ill.: Dryden Press.

Schlozman, K. L., & Tierney, J. T. (1986). *Organized Interests and American Democracy*. New York: Longman Higher Education.

Simon, H. A. (1953). Notes on the Observation and Measurement of Political Power. *Journal of Politics*, *15*(4), 500–516.

Sloof, R. (1998). *Game-Theoretic Models of the Political Influence of Interest Groups*. Boston: Kluwer.

Sloof, R., & van Winden, F. (2000). Show Them Your Teeth First! A Game-Theoretic Analysis of Lobbying and Pressure. *Public Choice*, *104*, 81–120.

Smith, M. P. (2008). All Access Points Are Not Created Equal: Explaining the Fate of Diffuse Interests in the EU. *British Journal of Politics & International Relations*, *10*(1), 64–83.

Steinberg, D. A., & Shih, V. C. (2012). Interest Group Influence in Authoritarian States: The Political Determinants of Chinese Exchange Rate Policy. *Comparative Political Studies*, *45*(11), 1405–1434.

Stigler, G. J. (1971). The Theory of Economic Regulation. *The Bell Journal of Economics and Management Science*, *2*(1), 3–21.

Thomas, C. S. (Ed.). (2004). *Research Guide to U.S. and International Interest Groups*. Westport: Praeger.

Truman, D. B. (1971). *The Governmental Process. Political Interests and Public Opinion* (2nd ed.). New York: Knopf.

van Winden, F. (2008). Interest Group Behavior and Influence. In C. K. Rowley & F. Schneider (Eds.), *Readings in Public Choice and Constitutional Political Economy*. New York: Springer.

Victor, J. N. (2007). Strategic Lobbying. Demonstrating How Legislative Context Affects Interest Groups' Lobbying Tactics. *American Politics Research, 35*(6), 826–845.

Ward, H. (2004). Pressure Politics: A Game-Theoretical Investigation of Lobbying and the Measurement of Power. *Journal of Theoretical Politics, 16*(1), 31–52.

Wright, J. R. (1996). *Interest Groups & Congress*. Needham Heights, MA: Allyn & Bacon.

CHAPTER 3

Process Uncertainty: Political Decision-Making

Abstract In this chapter, I discuss approaches to understand political bargaining. I show that bargaining implies a conflict of interest which creates uncertainty over the expected policy outcome. I discuss how the political process in Germany is structured and show that process uncertainty depends on the type of political process. This uncertainty is consequential for interest group activities.

Keywords Bargaining · Political process · Process uncertainty
Bargaining model

One aspect of the literature on interest groups is that political processes which follow the influence attempt of the interest group receive little attention. Maybe this is mostly visible in the case of signaling models where the interest group is able to predict the outcome of its lobbying activity. The models short-circuit the political process. While the short-circuiting of politics is one extreme, other approaches like Gais and Walker (2001) completely abstract from political processes. Here, all that matters for interest groups is survival. In its extremity, this would imply that it does not matter what the group achieves—as long as it suffices to guarantee survival.

Both short-circuiting and the separation of lobbying from political processes creates the danger of misperceptions. In the real world, interest groups operate in an inherently uncertain environment. Public policies are determined in a bargaining process between political actors whose exact identity varies across political systems. In Germany, the relevant actors are usually members of the government, the party factions in parliament,

© The Author(s) 2019   41
S. Koehler, *Lobbying, Political Uncertainty and Policy Outcomes*,
https://doi.org/10.1007/978-3-319-97055-4_3

the parliamentary committees as well as the representatives of the States (Länder) in the Bundesrat.

The ultimate goal of interest groups is to influence outcomes. Influencing politicians is a means to an end. The decision to lobby depends therefore on the identity and powers of the decision-makers. The importance and identity of the political actors varies considerably by issue area and bill. This has consequences for the strategies interest groups choose to influence policies. In the following, I discuss some determinants of political bargaining in Germany, resulting uncertainties, and potential targets from an interest group's point of view.

3.1   Law-Making and Political Bargaining

All modern democracies are organized as representative systems (Lijphart 1999). Usually, they comprise three branches, the judicial branch, the executive branch and the legislative branch. One of the major aspects of legislatures is, whether they are unicameral or bicameral. The German system constitutes a bicameral system. The lower house is the *Bundestag* (BT). Its members are elected in an electoral system characterized by mixed-member proportional representation. The upper house is the *Bundesrat* (BR), which is constituted by representatives of the Federal states (*Länder*).

3.1.1   The Legislative Process in Germany

The Bundestag is the more important of the two chambers in Germany. One of the peculiarities of the German system is that only part of the bills are subject to approval by the Bundesrat. These types of bills are called 'Zustimmungsgesetz'. In about 60% of the cases the Bundestag can decide without the explicit consent of the Bundesrat. In these cases, the Bundesrat has the power to delay but not to stop legislation (Ismayr 2008). These types of bills are called 'Einspruchsgesetz'.

Tsebelis (2002) defines a veto player as a political actor whose consent is required to change policy. In this sense, the Bundesrat is a veto player in some cases but not in others. This creates different incentive structures for lobbying as the political process looks very different under the two procedures. From the point of view of an interest group, the level of process uncertainty is higher in the case of a bill which is subject to approval by the Bundesrat.

A bill can be introduced in three different ways in the German system. The most common type of initiative is a government initiative. A second possibility is that a party faction or a group of at least 5% of the members of parliament introduce a bill. The last option is an initiative originating in the Bundesrat, usually based on the initiative of one or more states (Rudzio 2011).

When the government introduces a bill, the first (mandatory) step is to forward it to the Bundesrat. Within a period of six weeks, the Bundesrat issues comments on the bill. Upon reception of the comments or after six weeks if no comments were issued by the Bundesrat, the government is allowed to introduce the bill (including the comment where applicable) in the Bundestag. Once introduced, the bill is referred to one or more committees. The committees have considerable influence on the content of the bill and advise the floor regarding the acceptance of the bill. If the bill is accepted on the floor, it is sent to the Bundesrat. The Bundesrat may accept the bill as is, in which case it becomes law after the first reading.

If the Bundesrat amends the bill, it is sent back to the Bundestag for the second reading. The Bundestag may accept the bill as is, in which case it becomes law. Otherwise the Bundestag may propose amendments. A third reading may follow if the amended bill is not accepted by the Bundesrat after the second reading. If the Bundesrat proposes amendments after the third reading, a conciliation committee with representatives of the Bundestag and the Bundesrat may be established. The conciliation committee tries to negotiate a compromise. The compromise is then introduced again in both chambers. The compromise is a take-it-or-leave-it offer and cannot be amended. If both chambers accept the compromise it becomes law. Otherwise the bill is formally rejected.

The description of the law-making process is based on the case where the bill is subject to approval by the Bundesrat (*Zustimmungsgesetz*). Under this procedure the Bundesrat can veto any law. A rejection ends the legislative process in any stage of the process. If the bill is not subject to approval (*Einspruchsgesetz*), a veto by the Bundesrat may be overridden by the Bundestag with an absolute majority (Ismayr 2008). The Bundestag has absolute veto power in all cases. As such the Bundestag is a more important target for interest groups which is also reflected in the significantly higher number of network ties found by Knoke et al. (1996).

In the case of an initiative originating in the Bundesrat, the bill has to be handed to the government. The government may comment on the bill before it is introduced in the Bundestag. The rest of the legislative procedure follows the steps explicated above. Bills originating as initiatives by the parliamentary factions or a coalition of parliamentarians are introduced directly in the Bundestag without forwarding it to the Bundesrat for comments. The process starts immediately with the first reading in the Bundestag. Governments sometimes use this route to fast-track legislation, as this prevents delays caused by the Bundesrat (Ismayr 2008). The legislative procedures which follow are identical to the ones already described.

One of the peculiarities of the German system is that the implementation of laws is the responsibility of the German states (*Länder*). Detailed regulation is needed to ensure a common implementation in all German states. The federal state is therefore dependent inputs such as impact assessments by the Länder (Ismayr 2008). The increased complexity of legislation results in an increase in the demand for information which is partly provided by interest groups.

In Germany, there is a strong focus on corporate actors of interest representation stemming from corporatist structures. Corporatism is historically the predominant mode of interest representation in Germany (Sebaldt 2007). The German Bundestag, for example, runs a voluntary register for lobbyists.[1] Only associations are allowed (sic!) to register. Other important actors which engage in lobbying—like companies or consulting firms—do not meet the definition of the list and are not counted as interest groups in the sense of the list.

Streeck (1983) argues that the German interest groups (especially business associations and trade unions) are between the poles of collaborating with the state (and thereby fulfilling public functions) and trying to actively promote their individual interests. This tension between a corporatist mode of access and a pluralist self-consciousness determines the actions of interest groups. The state is able to partially absorb interest groups because this absorption also creates the valuable resources for the interest groups which they need for their survival. The corporatist mode is mainly relevant for the access of interest groups in the political system. Once access is given, the incentives for strategic *communication* are similar to political systems with less corporatist structures. Streeck also highlights the mix of pluralist and corporatist structures in Germany. When it comes to represent their interests, interest groups define themselves as pluralist actors and act accordingly. The logic of interest groups as entrepreneurs with two audiences was also highlighted by Ainsworth (1997).

[1] The register can be accessed at http://www.bundestag.de/dokumente/lobbyliste/.

Independent of group type, communication is major component of Lobbying activity. However, interest groups engage in other activities like mobilizing constituencies, encouraging members to vote, running public campaigns or donating to politicians' campaigns. They also provide benefits to members and in some countries help to implement policies, which is the case in countries with corporatist structures such as Germany (Streeck 1983) although the importance of these structures has declined considerably (Wehrmann 2007).

3.1.2   Bicameralism and Models of Bargaining

The legislative process creates uncertainty over the outcomes. One way to understand the bargaining (and thus possible outcomes) in a bicameral setup is the application of the spatial model of politics as Riker (1992) and Tsebelis and Money (1997) have shown. Tsebelis later broadened his approach and called it veto player theory (Tsebelis 2002). In addition to the spatial model, Tsebelis and Money (1997) also use procedural bargaining models to analyze inter-cameral bargaining from a game theoretic perspective.

In *Bargaining situations*, the parties which are bargaining share a common interest, so they have an incentive to cooperate. However, they disagree over the exact form of cooperation (Muthoo 1999). In legislative bargaining there are often policies which both chambers prefer over the Status Quo, yet, conflict arises over which policy to choose. In order to get to this problem theoretically, it is necessary to identify the relevant actors and spatially map their preferences. This allows to derive a prediction about the location and size of the set of policies which are preferred by both chambers over the Status Quo. This set of policies is called a winset (Tsebelis 2002).

Crucial in the spatial model is the idea that the position of the two chambers can be modeled by the position of the median legislator (Black 1958). The chambers can thus be treated as unitary actors. In these situations both chambers have incentives to cooperate by drafting and passing new legislation. However, there will be conflict about how far reaching the changes should be and in which direction legislation should be changed, i.e. which point to pick in the winset.

Tsebelis and Money (1997) analyze a version of this spatial model. They derive a set prediction, i.e. their model identifies a set of policies which are likely outcomes of the bargaining process (the winset). They capture the cooperation aspect of the bargaining problem, but only partly solve the conflicting part. Their model helps to establish whether cooperation is possible, but not necessarily how the conflict of interest will be solved. For this they would have to predict a single outcome.

One useful distinction is the argument that bicameralism has two dimensions. An *efficient* dimension and a *political* dimension. The distinction is inspired by the idea of *efficient* and *redistributive* dimensions defined by Tsebelis (1991). Efficiency is defined as the cooperative part, i.e. the aspects which are not conflicting. The *political* dimension is redistributive in the sense of a classical zero sum game. Gains of one actor are losses of the other actor.

Tsebelis and Money (1997) show that bicameralism reduces the conflict to the contract line between the two chamber's ideal points. The theory is powerful as it narrows down the predictions which would be made on the basis of the winset alone. However, it is not possible to predict policy outcomes with the help of the theory. All that can be said is that the policy should be located on the bargaining line between the two chambers. If we observe a certain policy on that line, the theory cannot explain how this particular policy was chosen. In fact, this indeterminacy of the exact point at which agreement occurs in barter has been long known (Edgeworth 1881).

In order to get a more precise theoretical prediction, one has to impose more structure on the problem. A first approach was John Nash's Bargaining Solution (Nash 1950). Nash derives a unique solution from a set of axioms which he postulates rational decision-makers would adhere to. This axiomatic approach has been criticized for some of these axioms used to determine the unique solution of the bargaining problem. Alternatives based on the different sets of axioms have successively been proposed (e.g. Kalai and Smorodinsky 1975).

Axiomatic representations of the bargaining situation ignore the bargaining process (Osborne and Rubinstein 1990). The approach is static, while the problem is dfynamic. In response to the critique, procedural models have been proposed. A pioneer was Ståhl (1972) who analyzed an alternating offer game within a finite horizon setting. In this game, the players alternate in making a proposal until one of the players accepts a proposal made by the other. Based on the concept of time preference (Fishburn and Rubinstein 1982) and Ståhl's model, Rubinstein (1982) developed his famous infinite horizon model of bargaining. He demonstrates that the model has a unique solution and agreement is immediate in the case of complete information as bargaining is costly. The first mover in the model has an advantage over the second mover.

If one restricts some parameters, the Nash bargaining solution can be derived from a procedural bargaining model which justifies the use in many cases in economics (Binmore et al. 1986) or political science (Powell 2002). Versions of the Nash bargaining solution have for example been used to analyze legislative decision-making in the European Union (Thomson et al. 2006).

Several authors have shown that the players in a dynamic bargaining game will not reach immediate agreement if there is uncertainty about the preferences of others (Rubinstein 1985) or asymmetric information about the value of the stakes which are to be distributed (Grossman and Perry 1986). In both cases, the players are willing to incur the costs of bargaining as the process discloses information to the uninformed players. Tsebelis and Money (1997) show empirically that the level of uncertainty influences the length of bargaining.

A generalization of the Rubinstein bargaining model widely used in political science is the model by Baron and Ferejohn (1989). They develop a procedural bargaining model which has been used and adopted to describe decision-making in the US Congress (Ansolabehere et al. 2003; Baron and Ferejohn 1989), coalition formation (Baron 1989) or competitive lobbying in majority-rule institutions (Baron 2006). The first mover advantage is also a feature of the Baron–Ferejohn model.

A much simpler model was developed by Romer and Rosenthal (1978) based on the spatial model of politics. Their model is the basis for simple veto bargaining models. The Romer–Rosenthal model can be seen as a special case of the Rubinstein–Ståhl model where many parameters are restricted. The strongest restriction is on the temporal dimension. In this model, an agenda-setter makes a take-it-or-leave-it proposal to another actor. If the other actor accepts there is an agreement. Otherwise, the Status Quo remains (or the reversion point is realized, if different from the Status Quo). That is, bargaining is essentially restricted to one period. The model has many features which are present in the Rubinstein model, for example, the advantage of the agenda-setter. Models of this type have been fruitfully used to analyze interactions of parliaments and presidents (Matthews 1989; Cameron 2000; McCarty 2000; Cameron and McCarty 2004) or bicameral legislatures like US Congress and other (Krehbiel 1998; Rogers 1998; Cutrone and McCarty 2009). Tsebelis (2002) uses a similar argument to determine the outcome of interinstitutional bargaining.

Procedural models of decision-making based on the Romer-Rosenthal ideas were developed to theoretically analyze decision-making in the European Union (e.g. Stokman and Bueno de Mesquita 1994; Garrett 1995; Crombez 1996, 2000, 2001; Steunenberg 1994, 1996), and empirically (e.g. Steunenberg and Selck 2006). All these models provide insights into how individual characteristics (player's discount factors and preferences) and institutional features (veto/proposal rights, procedural rules) determine bargaining power and outcomes.

The models apply game theoretic tools to the interaction of the three major institutional actors in the European Union: The Commission, the European Parliament, and the Council. The common point of departure is the identification of the relevant actors and the translation of the procedural rules established by the treaties (Cooperation, Consultation, and Co-decision procedure) into stylized procedural models (e.g. Tsebelis and Yataganas 2002). Often are institutional actors treated as unitary actors. Sometimes these institutional actors are further decomposed into partisan actors like the countries in the Council (Crombez 2000). The preferences of these actors are usually assumed as exogenously given. Here lies the potential for linking interest representation and decision-making (Crombez 2002).

Procedural models model the decision-making process as a strategic interaction of rational actors that are constrained by procedural rules. They are spatial models where actors preferences are modeled as positions in a policy space. Several aspects play a crucial role for the balance of power between the actors and the understanding of the dynamics of the decision-making process. The first one is gatekeeping. An actor who decides about which issues are put on the agenda has gatekeeping power (Crombez et al. 2006).

Another source of power is agenda-setting power. An actor who has the ability to make a proposal that is hard (or impossible) to amend has a similar argument can be made with respect to the influence of actors in the decision-making process (Selck and Steunenberg 2004) agenda-setting power (Tsebelis and Garrett 1996).

The last concept is veto power (Tsebelis 2002; Tsebelis and Garrett 2001). Actors that have to agree on a change of the Status Quo have veto power as they can block the decision with a no vote. All of these concepts are in varying degree reflected in procedural models of decision-making. The powers arise from the rules of the game, i.e. formal institutional provisions.

In the European context, these are the various treaties that establish the various decision making procedures.

One of the most important features of the procedural models is that the agenda-setter is constrained in his choice by the veto power of the other actor(s). Given the preferences of the political actors, the outcome of the model is unique in a complete information environment. The agenda-setter is quite powerful as 'it makes a difference where a bill is introduced first.' (Tsebelis and Money 1997, 101).

Rogers (1998) therefore goes one step further. He analyzes the choice of the bargaining sequence as a strategic decision of the actors in an environment with incomplete information. In his model, the chambers' interaction is driven by the distribution of information about the state of the world. However, to be informed or not is a decision solely made by the chambers themselves. Once the chamber pays a cost, it can become perfectly informed about the state of the world. I will add another layer here by explicitly modeling the chambers' dependence on biased information provided by interest groups. However, I drop the idea of an endogenous sequence.

The bargaining problem is a situation, in which coordinated action can produce mutually beneficial outcomes for all participants, but there is conflict about how exactly the gains from cooperation should be divided among the bargaining parties (Schellling 1960, Chapter 2). Bicameralism can be seen as an institutional setup which creates the need for bargaining in the first place—provided the chambers have conflicting preferences (Heller 2007).

In many instances, the preferences of the two chambers are conflicting even in situations of unified government. Where do the conflicts of interest between the chambers come from? In most cases, the chambers in bicameral legislatures represent distinct constituencies (Russel 2001). An example of this is the German Bundestag which represents the population as a whole and the Bundesrat which represents the interests of the German states (*Länder*). Both represent distinct sets of interests, which are partially cross-cutting.

While the Federal State has strong powers of issuing legislation, the implementation is the responsibility of the individual states (Länder) (von Beyme 2010). Even if the same coalition of parties dominates in the Bundestag and the Bundesrat there may be disagreement between the chambers because of the costs for implementation and regional differences in economic structure and hence expected impacts of a bill.

From an interest group's point of view this constitutes both a chance and a problem: The bargaining process creates possibilities to influence the outcome by multiplying access points. At the same time, the bargaining between political actors decreases the predictability of outcomes. The result is *process uncertainty*, which results from reduced predictability of outcomes.

The conflict of interest between the chambers may be so dramatic that the second chamber vetoes a bill (or threatens to do so). From an interest group's perspective this creates a second layer of uncertainty: The bill may be voted down after the lobbying activity in which case the Status Quo would prevail even if the interest group was able to influence one or some of the decision-makers involved in the bargaining.

3.1.3   The Coalition Game

The political uncertainty created for interest groups by the bicameral structure has been explained. But what about the cases where the law is not subject to approval by the Bundesrat? Coalition bargaining entails process uncertainty but it differs from the process uncertainty inherent in the bicameral game.

Parliamentary systems are usually characterized by strong party discipline and dominant governments (Lijphart 1999). From an interest group's perspective this dramatically reduces the likelihood of a veto. The question is not so much whether the bill will become law, it is rather what it will look like. In Germany, the percentage of government bills which became law is significantly higher than for opposition bills or bills originating in the Bundesrat (Ismayr 2012).

Governments in parliamentary systems with a proportional electoral system tend to produce multiparty legislatures. This point is known as Duverger's law Duverger (1951). Multiparty legislatures in parliamentary systems tend to have coalition governments. Coalition governments are not unitary actors. The structural similarity between a bicameral legislature and a coalition government of two parties when represented by a spatial model is striking. Instead of inter-cameral bargaining, the policy is the result of bargaining between the coalition partners.

Martin and Vanberg (2004), for example, analyze potential conflict within coalition governments. Their argument builds on agency theory. While the coalition partners may agree on a coalition compromise, the formulation of a bill is usually delegated to a (partisan) minister. This creates a principal agent problem. This problem rests on the fundamental uncertainty inherent in the situation. A minister's ability to monitor the fellow minister's compliance with the coalition compromise usually depends on information about (a) the Feasibility of policies in the respective policy area (the choice set) and (b) the resulting policy outcomes. Interest groups provide both types of information information in order to influence the outcome of the coalition game. They may sound 'fire alarms' (McCubbins and Schwartz 1984). As such, the informational dependencies in coalition governments are quite similar to the dependencies found in bicameral bargaining.

Of course, the provision of information by interest groups is not the only possibility. Coalition partners use a range of strategies to control each other. Examples of institutional arrangements are parliamentary scrutiny (Martin and Vanberg 2004, 2005) or junior ministers (Thies 2001). Due to the constant struggle about policy outcomes the process is quite uncertain from an interest groups point of view.

Political (or process) uncertainty plays a role in several models. For example, Huber and McCarty (2001) analyze how the interaction of political uncertainty and cabinet decision-making rules shapes bargaining outcomes in coalition governments. This aspect is particularly relevant for the German case. They model the internal organization of coalition governments. They show that the right to request a vote of confidence is a strong predictor of the power to extract concessions from the coalition partner. These institutional features will therefore lead to deviations from the coalition compromise.

3.2   Process Uncertainty

I have argued that *process uncertainty* can be defined as the ex-ante possible variation in outcomes, i.e. the length of bargaining line. Three factors determine process uncertainty: (1) The degree of fundamental uncertainty, (2) the potential for policy change (the conflict of interest), and (3) the bargaining procedure.

First, fundamental uncertainty feeds back into the degree of political uncertainty, because the political actors have different incentives when uncertainty differs. As long as uncertainty is not too high, they are better-off changing the policy, even not obtaining their optimal outcomes in the *absence* of lobbying.

Second, the potential for policy change (or the room for maneuver) is given by the bargaining line between the chambers. The more conservative chamber is effectively setting a restriction and therefore decisive for the political room for maneuver. Conservative is understood here not in ideological terms, but in relation to the desired move away from the Status Quo. The smaller this distance, the more 'conservative' the chamber. The exact room for maneuver is determined by the distribution of the Status Quo and the ideal points of the two chambers or coalition partners.

Third, political uncertainty results from differences in processes. Processes may be highly structured, as is true for most legislative processes. Or they may be rather unstructured as is the case for coalition bargaining in the government. In either case there is considerable leeway. A bill may be changed in committee or it may be voted down in the second chamber. It may be blocked by a coalition partner or dragged into the direction of one of the other. Likewise, a typical legislative process in a parliamentary system with a high degree of party discipline reduces political uncertainty. The less disciplined interaction in the US Congress ceteris paribus leads to higher political uncertainty.

REFERENCES

Ainsworth, S. H. (1997). The Role of Legislators in the Determination of Interest Group Influence. *Legislative Studies Quarterly, 22*(4), 517–533.

Ansolabehere, S., Snyder, J. M., & Ting, M. M. (2003). Bargaining in Bicameral Legislatures: When and Why Does Malapportionment Matter? *The American Political Science Review, 97*(3), 471–481.

Baron, D. P. (1989). A Noncooperative Theory of Legislative Coalitions. *American Journal of Political Science, 33*(4), 1048–1084.

Baron, D. P. (2006). Competitive Lobbying and Supermajorities in a Majority-Rule Institution. *Scandinavian Journal of Economics, 108*(4), 607–642.

Baron, D. P., & Ferejohn, J. A. (1989). Bargaining in Legislatures. *American Political Science Review, 83*(4), 1181–1206.

Binmore, K., Rubinstein, A., & Wolinsky, A. (1986). The Nash Bargaining Solution in Economic Modeling. *Rand Journal of Economics, 17*(2), 176–188.

Black, D. (1958). *The Theory of Committees and Elections.* London: Cambridge University Press.

Cameron, C. (2000). *Veto Bargaining.* Cambridge: Cambridge University Press.

Cameron, C., & McCarty, N. (2004). Models of Vetoes and Veto Bargaining. *Annual Review of Political Science, 7*(1), 409–435.

Crombez, C. (1996). Legislative Procedures in the European Community. *British Journal of Political Science*, *26*, 199–228.

Crombez, C. (2000). Institutional Reform and Co-decision in the European Union. *Constitutional Political Economy*, *11*(1), 41–57.

Crombez, C. (2001). The Treaty of Amsterdam and the Codecision Procedure in the European Union. In G. Schneider & M. Aspinwall (Eds.), *The Rules of Integration. The Institutionalist Approach to European Studies*. Manchester: Manchester University Press.

Crombez, C. (2002). Information, Lobbying and the Legislative Process in the European Union. *European Union Politics*, *3*(1), 7–32.

Crombez, C., Groseclose, T., & Krehbiel, K. (2006). Gatekeeping. *Journal of Politics*, *68*(2), 322–334.

Cutrone, M., & McCarty, N. (2009). Does Bicameralism Matter? In B. R. Weingast & D. A. Wittman (Eds.), *Oxford Handbook of Political Economy*, Chapter 10 (pp. 180–195). Oxford: Oxford University Press.

Duverger, M. (1951). *Political Parties: Their Organization and Activity in the Modern State*. New York: Wiley.

Edgeworth, F. Y. (1881). *Mathematical Psychics*. London: Kegan Paul.

Fishburn, P. C., & Rubinstein, A. (1982). Time Preference. *International Economic Review*, *23*(3), 677–694.

Gais, T. L., & Walker, J. L. (2001). Pathways to Influence in American Politics. In J. L. Walker (Ed.), *Mobilizing Interest Groups in America*, Chapter 6 (pp. 103–121). Ann Arbor: University of Michigan Press.

Garrett, G. (1995). From the Luxembourg Compromise to Codecision: Decision Making in the European Union. *Electoral Studies*, *14*(3), 289–308.

Grossman, S. J., & Perry, M. (1986). Sequential Bargaining Under Asymmetric Information. *Journal of Economic Theory*, *39*(1), 120–154.

Heller, W. B. (2007). Divided Politics: Bicameralism, Parties, and Policy in Democratic Legislatures. *Annual Review of Political Science*, *10*(1), 245–269.

Huber, J. D., & McCarty, N. (2001). Cabinet Decision Rules and Political Uncertainty in Parliamentary Bargaining. *The American Political Science Review*, *95*(2), 345–360.

Ismayr, W. (2008). Gesetzgebung im politischen System Deutschlands. In W. Ismayr (Ed.), *Gesetzgebung in Westeuropa: EU-Staaten und Europäische Union* (pp. 383–430). Wiesbaden: VS Verlag für Sozialwissenschaften.

Ismayr, W. (2012). *Der Deutsche Bundestag* (3rd ed.). Wiesbaden: Springer VS.

Kalai, E., & Smorodinsky, M. (1975). Other Solutions to Nash's Bargaining Problem. *Econometrica*, *43*(3), 513–518.

Knoke, D., Pappi, F. U., Broadbent, J., & Tsujinaka, Y. (1996). *Comparing Policy Networks*. Cambridge: Cambridge University Press.

Krehbiel, K. (1998). *Pivotal Politics: A Theory of U.S. Lawmaking*. Chicago: University of Chicago Press.

Lijphart, A. (1999). *Patterns of Democracy: Government Forms and Performance in Thirty-Six Countries*. Yale: Yale University Press.

Martin, L. W., & Vanberg, G. (2004). Policing the Bargain: Coalition Government and Parliamentary Scrutiny. *American Journal of Political Science, 48*(1), 13–27.

Martin, L. W., & Vanberg, G. (2005). Coalition Policy Making and Legislative Review. *The American Political Science Review, 99*(1), 93–106.

Matthews, S. A. (1989). Veto Threats: Rhetoric in a Bargaining Game. *The Quarterly Journal of Economics, 104*(2), 347–369.

McCarty, N. (2000). Proposal Rights, Veto Rights, and Political Bargaining. *American Journal of Political Science, 44*(3), 506–522.

McCubbins, M., & Schwartz, T. (1984). Congressional Oversight Overlooked: Police Patrols Versus Fire Alarms. *American Journal of Political Science, 28*(1), 16–79.

Muthoo, A. (1999). *Bargaining Theory with Applications*. Cambridge: Cambridge University Press.

Nash, J. F. (1950). The Bargaining Solution. *Econometrica, 18*(2), 155–162.

Osborne, M. J., & Rubinstein, A. (1990). *Bargaining and Markets*. Emerald: Bingley.

Powell, R. (2002). Bargaining Theory and International Conflict. *Annual Review of Political Science, 5*, 1–30.

Riker, W. H. (1992). The Justification of Bicameralism. *Revue internationale de science politique* [International Political Science Review], *13*(1), 101–116.

Rogers, J. R. (1998). Bicameral Sequence: Theory and State Legislative Evidence. *American Journal of Political Science, 42*(2), 1025–1060.

Romer, T., & Rosenthal, H. (1978). Political Resource Allocation, Controlled Agendas, and the Status Quo. *Public Choice, 33*(4), 27–43.

Rubinstein, A. (1982). Perfect Equilibrium in a Bargaining Model. *Econometrica, 50*(1), 97–109.

Rubinstein, A. (1985). A Bargaining Model with Incomplete Information About Time-Preferences. *Econometrica, 53*(5), 1151–1172.

Rudzio, W. (2011). *Das politische System der Bundesrepublik Deutschland* (8th ed.). Wiesbaden: VS - Verlag für Sozialwissenschaften.

Russel, M. (2001). What Are Second Chambers for? *Parliamentary Affairs, 54*, 442–458.

Schelling, T. C. (1960). *The Strategy of Conflict*. Cambridge, MA: Harvard University Press.

Sebaldt, M. (2007). Lobbying in Deutschland - Begriff und Trends. In R. Kleinfeld, A. Zimmer, & U. Willems (Eds.), *Lobbying - Strukturen. Akteure. Strategien* (pp. 92–94). Wiesbaden: Springer (VS).

Selck, T. J., & Steunenberg, B. (2004). Between Power and Luck: The European Parliament in the EU Legislative Process. *European Union Politics, 5*(1), 25–46.

Ståhl, I. (1972). *Bargaining Theory*. Stockholm: Economic Research Institute.

Steunenberg, B. (1994). Decision-Making Under Different Institutional Arrangements - Legislation by the European-Community. *Journal of Institutional and Theoretical Economics, 150*(4), 642–669.

Steunenberg, B. (1996). Agent Discretion, Regulatory Policymaking, and Different Institutional Arrangements. *Public Choice, 86*(3–4), 309–339.

Steunenberg, B., & Selck, T. J. (2006). Testing Procedural Models of EU Legislative Decision Making. In R. Thomson, F. N. Stokman, C. H. Achen, & T. König (Eds.), *The European Union Decides* (pp. 54–85). Cambridge: Cambridge University Press.

Stokman, F. N., & Bueno de Mesquita, B. (1994). *European Community Decision Making: Models, Applications, and Comparisons.* New Haven: Yale University Press.

Streeck, W. (1983). Between Pluralism and Corporatism: German Business Associations and the State. *Journal of Public Policy, 3*(3), 265–284.

Thies, M. F. (2001). Keeping Tabs on Partners: The Logic of Delegation in Coalition Governments. *American Journal of Political Science, 45*(3), 580–598.

Thomson, R., Stokman, F. N., Achen, C. H., & König, T. (Eds.). (2006). *The European Union Decides.* Cambridge: Cambridge University Press.

Tsebelis, G. (1991). *Nested Games: Rational Choice in Comparative Politics.* Berkeley: University of California Press.

Tsebelis, G. (2002). *Veto Players: How Political Institutions Work.* Princeton: Princeton University Press.

Tsebelis, G., & Garrett, G. (1996). Agenda Setting Power, Power Indices, and Decision-Making in the European Union. *International Review of Law and Economics, 16*(3), 345–361.

Tsebelis, G., & Garrett, G. (2001). The Institutional Foundations of Intergovernmentalism and Supranationalism in the European Union. *International Organization, 55*(2), 357–390.

Tsebelis, G., & Money, J. (1997). *Bicameralism.* Cambridge: Cambridge University Press.

Tsebelis, G., & Yataganas, X. (2002). Veto Players and Decision-Making in the EU After Nice: Policy Stability and Bureaucratic/Judicial Discretion. *Journal of Common Market Studies, 40*(2), 283–307.

von Beyme, K. (2010). *Das Politische System der Bundesrepublik Deutschland* (11th ed.). Wiesbaden: VS - Verlag für Sozialwissenschaften.

Wehrmann, I. (2007). Lobbying in Deutschland - Begriff und Trends. In R. Kleinfeld, A. Zimmer, & U. Willems (Eds.), *Lobbying - Strukturen. Akteure. Strategien* (pp. 36–64). Wiesbaden: Springer (VS).

CHAPTER 4

Modeling Interest Group Communication Strategies

Abstract In this chapter, I combine the insights from the literatures on lobbying and bargaining into a formal model of interest group activities. I distinguish public and private communication as classes of structurally equivalent lobbying activities. I show that the bargaining environment constrains lobbying activities. I demonstrate theoretically that the decision to send private messages depends on the ideological distance to the constraining chamber, while the decision to engage in public communication rests on the distance to the expected policy outcome.

Keywords Signaling model · Veto bargaining model · Lobbying strategy · Decision-making · Nash equilibrium

The chapter lays out the model of lobbying and decision-making in a bicameral political system. The model is not restricted to a bicameral setting. Any situation with two decision-makers will follow the same logic. For example, two parties in a coalition government in a unicameral setting will be structurally equivalent. The case is important for the empirical analysis. Note that the model is explicitly based on the idea of legislatures with strong party discipline, such as most European political systems. It is not applicable in cases with weak party discipline, such as the United States Congress.

© The Author(s) 2019
S. Koehler, *Lobbying, Political Uncertainty and Policy Outcomes,*
https://doi.org/10.1007/978-3-319-97055-4_4

4.1   THE MODEL SETUP

The model comprises three players. An interest group called G and the two chambers of a bicameral legislature. I call the two chambers Upper House (U) and Lower House (L), respectively. The two chambers decide on a policy in a one-dimensional policy space[1] which is represented by the real line. The Status Quo $Q \in \mathbb{R}$ is a point in this policy space, as is a policy $y \in \mathbb{R}$.

The policy outcome x is determined by the policy y and the realization of a discrete random variable θ according to $x = y + \theta$. The realizations of θ are indexed by j and determine the possible states of the world . I specifically assume that θ has only two realizations ($j \in \{1, 2\}$). The shocks are symmetric and $-\theta_1 = \theta_2$. The shocks are equally likely and occur with probability $\frac{1}{2}$. The expected value of θ is zero, i.e., the policy outcome is unbiased in expectation. I assume further, without loss of generality, that $\theta_1 < 0$. The prior distribution of θ is common knowledge.

The players have preferences over policy *outcomes*. I assume that all actors are risk averse. In particular, I assume that the preferences of the chambers can be modeled using a quadratic utility loss functions as

$$U_i(x) = -(x - X_i)^2$$

where X_i is the idealpoint of player i, $i \in \{l, u\}$. The ideal point is exogenously given. I do not make any assumptions about what determines the ideal point. Note, however, that the risk aversion has consequences. The decision-makers are ceteris paribus strictly better-off if they are informed about the shock.[2]

The utility functions which represent the interest group's preferences are similar, but entail an additional term which reflects the fact that lobbying is costly.

$$U_g(x) = -(x - X_g)^2 - I_g \cdot C_g$$

---

[1] Tsebelis and Money (1997, Chapter 3) have shown that bicameralism will reduce the conflict between the two chambers to one dimension, so this assumption is not purely to increase mathematical convenience, but it is also justifiable on theoretical grounds.

[2] To see this note that the chambers' induced preferences over policies $EU_i(y, \theta) = -(y + E[\theta] - X_i)^2 - var(\theta)$ contain the variance of the random variable θ. If they are informed, the variance term is zero. In all other cases, it is strictly positive. Note that the variance of θ is given as $var(\theta) = \sum_{i=1}^{2} \left(\frac{1}{2} \cdot \theta_i^2\right) - E[\theta]^2$. Substituting $-\theta_2$ for θ_1 this simplifies to θ_2^2. The expected utility can thus be written as $EU_i(y, \theta_2) = -(y - X_i)^2 - \theta_2^2$.

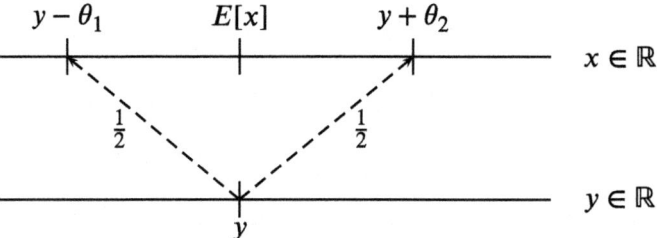

Fig. 4.1 Mapping from policy to outcome space

I_l is an indicator function that is one if the interest group decides to lobby (send a message) and zero otherwise. The cost $C_g \in \mathbb{R}$ depends on the strategy chosen, a point that will be explicated further when I analyze the full game. The interest group knows the State of the World. However, the policy outcome x is a random variable from the interest groups point of view as it is determined by bargaining of the two chambers.

I this setup, there is an analytical distinction between a policy and an outcome space. This distinction has been introduced by Gilligan and Krehbiel (1987) in their model of legislative decision-making. Figure 4.1 demonstrates graphically, how the policy space is mapped into an outcome space. Implementing the policy y yields the policy outcome $E[x] = y$ in expectation. Depending on the state of the world it will lead to either $y + \theta_1$ or $y + \theta_2$. This setup reflects the dependency of policymakers on information provided by better informed interest groups.

Figure 4.2 shows graphically the consequence of the policy shocks. In order to implement a policy x^* the decision-makers have to adjust the policy to the State of the World, which makes the optimal policy a function of the parameter θ. The chambers' have no knowledge about the exact State of the World. Their preferences are state dependent. In consequence, their induced preferences over policies depend negatively on the variance of θ. Lower variance is always better for the chambers.[3]

An important aspect of decision-making is the distribution of preferences and the Status Quo. To have a common vocabulary for defining a chambers' positions I resort to some standard definitions (e.g. Rogers

---

[3]To see this calculate the expected utility over the potential outcomes of policy y, which yields $-(y - X_i)^2 - var(\theta)$. See Kreps (1990, 87–91) for a discussion of induced preferences.

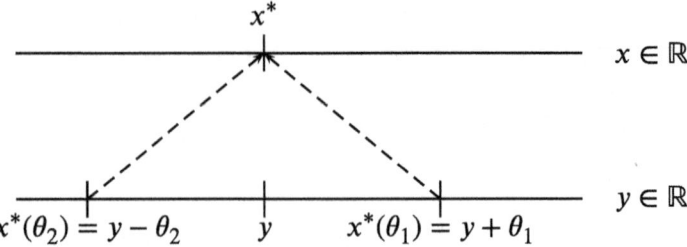

Fig. 4.2 Optimal policies as a function of θ

1998, 1036) and denote a chamber as *accommodating*, when the other house's ideal point is preferred by the chamber to the Status Quo. A chamber is *compromising*, when any movement from the Status Quo to its ideal outcome makes both chambers better-off, and as *recalcitrant*, when the Status Quo is located between the idealpoints of the chambers, so there is no policy preferred to the Status Quo which implies a move toward the other chamber's idealpoint.

Note, that if one chamber is accommodating, the other is necessarily compromising. When the chambers are recalcitrant, no policy change is possible as moving in any direction leaves one of the chambers worse-off. An alternative way to express this in the context of spatial models is that the Status Quo lies in the *core* (Tsebelis 2002). This situation is rather uninteresting as influential lobbying is impossible in this setup. We should not observe any lobbying as there will be no policy change. This is in line with the observation of a *Status Quo* bias in politics (Baumgartner et al. 2009) which is justified on a similar argument. However, Baumgartner et al. do not use a strategic model and do not analyze all implications of the observation.

Without loss of generality I assume $Q < X_u < X_l$. In this setting, the Upper House is compromising and the Lower House is accommodating. I assume further that the distance between the chambers is greater than the distance between the Upper House and the Status Quo, i.e. $|X_l - X_u| > |Q - X_u|$. This implies that the conflict between the chambers is relatively large. In particular, the Lower House would like to move the policy further than the Upper House is willing to accept. This assumption reduces the number of cases to be analyzed but does not influence the results of the model. The only difference is the relative strength of the restriction which

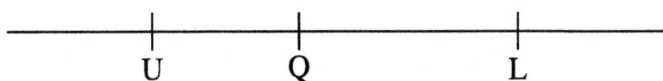

Fig. 4.3 Example for distribution of ideal points with recalcitrant chambers

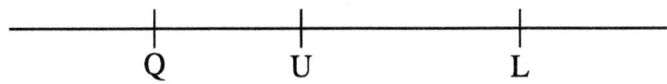

Fig. 4.4 Example for distribution of ideal points and SQ with compromising Upper House (U) and accommodating Lower House (L)

the compromising Upper House imposes on the accommodating Lower House.

Figure 4.3 shows an example for a distribution of ideal points and the Status Quo where the chambers are recalcitrant. Figure 4.4 shows an example of a distribution of preferences and the Status Quo where the House is compromising and the Lower House is accommodating. In addition, the assumption, that $|X_l - X_u| > |Q - X_u|$ holds in the setup in Fig. 4.4.

Let further $|X_u - Q| \geq \theta_2$, i.e. the size of the shock is smaller than the distance between the Upper House's ideal point and the status quo. This ensures that even an uniformed Upper House would want to change the policy. I assume that $X_l > 2X_u - Q + \theta_2$. The conflict of interest between the chamber is large, relative to the level of uncertainty.

Messages sent to only one chamber are *private*, the other chamber is not able to observe the message or its content. If both chambers are targeted the message is *public*, the message and its content are observed by both chambers.

No assumptions are made about the exact type of lobbying activity. A whole range of communication activities could be used to send such a message. A press statement in a newspaper, on Radio or on TV would fulfill this purpose. The participation in a hearing, the mobilization of members, or organizing a mass demonstration would be other examples. This is because the message costs ensure that interest groups only engage in informative communication. The costs for a public message are exogenously given and are common knowledge. The costs for a private message are also exogenously given and common knowledge.

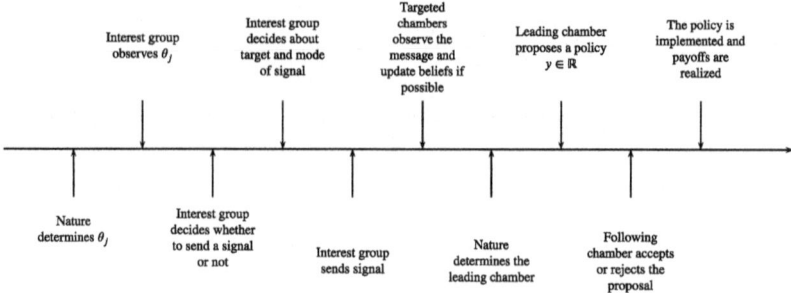

Fig. 4.5 Timing of the game

The message space is given as M = {m, n} where m denotes the message sent in equilibrium and n denotes no message in equilibrium. A signaling strategy for the interest group is given by $\sigma_g = \theta \times \mathbb{R} \to M$.

The game proceeds as shown in Fig. 4.5. Nature chooses the realization of θ. The Interest group observes the true value of θ. Then the interest group decides whether to send a message or not. If it decides to send a message it pays the cost corresponding to the target and privacy level. The targeted chambers observe the message (or its absence) and update their beliefs about the true value of θ if possible. Following the lobbying stage, one of the chambers is randomly recognized as the agenda-setter and makes a take-it-or-leave-it proposal to the other chamber. I assume that the recognition probability is $p = \frac{1}{2}$. The second chamber can either accept or veto the policy proposal. Acceptance leads to the implementation of the policy while in case of a veto the Status Quo prevails. Then the game ends and the payoffs are realized.

The solution concept for the model is Perfect Bayesian Equilibrium with the elimination of strictly dominated strategies. In equilibrium, all actors choose a strategy which maximizes their expected utility and update their beliefs according to Bayes' Rule whenever possible. In the following, I discuss the model.

4.2   The Bargaining Stage

In order to model the bargaining between the two chambers, a veto bargaining model in the tradition of Romer and Rosenthal (1978) is used. It bears similarities to the models of Matthews (1989) and Rogers (1998).

However, the role of the proposer is fixed in Matthews (1989) and endogenous in Rogers (1998).

I assume that each of the chambers is recognized as the proposer with probability $p = \frac{1}{2}$. The assumption of random recognition may seem odd, but it is, for example, used in the seminal bargaining model of Baron and Ferejohn (1989). The assumption of random recognition is a way to model the uncertainty an interest group faces about the outcome of the political process. Models are "purpose relative" Morton (1993). The purpose of my model specification is to describe the process uncertainty emanating from the bargaining situation and its effect on interest group behavior in the simplest possible way.

In Germany and many other bicameral systems, bills can be introduced in either of the two chambers of the legislature. In case of a coalition government, either of the coalition partners might pull the policy closer to its ideal point. Both types of uncertainty generators are captured with this model specification. What matters for the choice of lobbying strategy is the interest group's preferences over lotteries of outcomes, a point also raised in (Esterling 2004). The model is therefore more useful to explain actual behavior of interest groups than most signaling models of lobbying.

The chambers' actions depend on the prior moves in the game. Once the interest group has moved, four scenarios are possible. They are characterized by the information distribution induced by the interest group's action. Denote an informed chamber I and an uninformed NI. The four possible scenarios are: (I,I), (I,NI), (NI,I), and (NI,NI). That is, either both chambers are informed (after a public message), one of them (after a private message) or none. The (I,NI) and (NI,I) scenarios are characterized by an informational asymmetry between the two chambers. The (I,NI) scenario denotes a situation where the compromising chamber (the Upper House) is informed. In the (NI,I) scenario, the accommodating chamber (the Lower House) is informed. Note that in an (I,NI) and (NI,I) scenario, the chambers only know their own information state.

Both chambers will accept a proposal if the expected utility from the new policy is higher than the utility derived from the Status Quo. They will reject a proposal if the expected utility from the new policy is lower than the utility derived from the Status Quo. A voting strategy at this stage is a mapping $\sigma_{vi} : \mathbb{R} \to \{0, 1\}$, were 1 denotes acceptance of the proposal and 0 denotes a veto. For chamber $i \in \{u, l\}$ the veto strategy is therefore given by

$$\sigma_{vi} = \begin{cases} 1 & \text{if } EU_i(y, \theta) \geq U_i(Q) \\ 0 & \text{if } EU_i(y, \theta) < U_i(Q) \end{cases} \tag{4.1}$$

A proposal strategy for chamber i, $i \in \{u, l\}$ in the bargaining stage is a mapping $\sigma_{pi} : M \rightarrow \mathbb{R}$. Proposition 1 summarizes the equilibrium proposal strategies of the two chambers. For easier comparison the equilibrium proposals are also given in Table 4.1. Figure 4.6 shows the equilibrium outcomes and their relative frequency. The equilibrium proposal is denoted y^*.

Information about the value of θ has to be communicated via the bills. That is, an uninformed chamber has to infer the exact *State of the World* from the content of the bill whenever possible. In the case that the compromising Upper House is informed, it can signal the *State of the World* to the accommodating Lower House. Signaling from the Lower House to the Upper House is not possible as the Lower House has incentives to deceive the House when $\theta = \theta_2$.

Proposition 1 *The following strategies constitute an equilibrium at the legislative stage*

(a) *In the (I,I) scenario, where both chambers know their type, if the compromising Upper House is recognized, the two types separate in equilibrium and propose $y^* = X_u - \theta_j$, with $j \in \{1, 2\}$ corresponding to the type. Both types of the Lower House accept this offer. If the accommodating Lower House is recognized the types separate and propose $y^* = 2X_u - Q - \theta_j$, with $j \in \{1, 2\}$ corresponding to the type. The two types of the Upper House accept the proposal.*

(b) *In the (I,NI) scenario, where the Upper House knows its type but the Lower House does not, the types of the House separate and propose $y^* = X_u - \theta_j$, if recognized. Both types of Lower House accept this proposal. If the Lower*

Table 4.1 Equilibrium proposals depending on information distribution for $|X_u - Q| \geq \theta_2$

Scenario	Upper House	Lower House
(I,I)	$X_u - \theta_j$	$2X_u - Q - \theta_j$
(I,NI)	$X_u - \theta_j$	$2X_u - Q - \theta_2$
(NI,NI)	$X_u - E[\theta]$	$X_u + \sqrt{(X_u - Q)^2 - \theta_2^2}$
(NI,I)	$X_u - E[\theta]$	$X_u + \sqrt{(X_u - Q)^2 - \theta_2^2}$

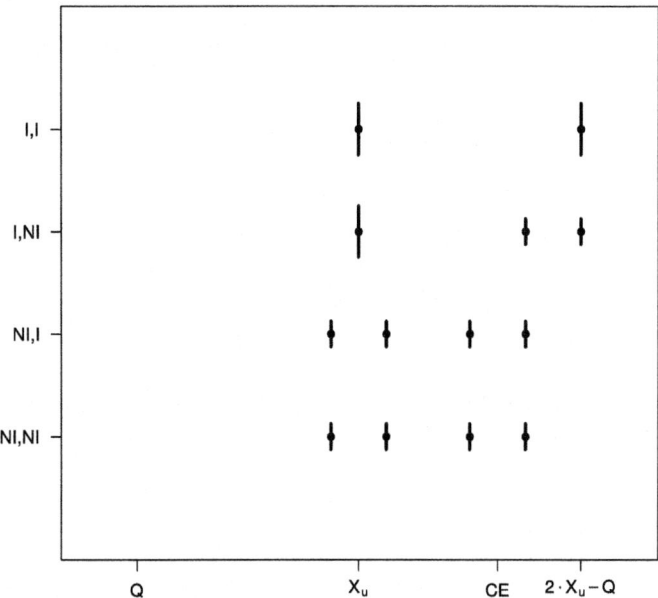

Fig. 4.6 Bargaining outcomes in the different possible scenarios. Dots denote outcomes, length of bar indicates relative frequency of the outcome

House is recognized the types pool and propose $y^* = 2X_u - Q - \theta_2$. Both types of the Upper House accept this proposal.

(c) *In the (NI,NI) scenario, where no chamber knows its type, the types of the Upper House pool when recognized and propose $y^* = X_u - E[\theta] = X_u$, which both types of the Lower House accept. If the Lower House is recognized the two types pool and propose $y^* = X_u + \sqrt{(X_u - Q)^2 - \theta_2^2}$, which both types of the Upper House accept.*

(d) *In the (NI,I) scenario, where the Lower House knows its type and the Upper House does not, the two types of the Upper House pool and propose $y^* = X_u - E[\theta] = X_u$ when recognized. The two types of the Lower House accept this proposal. When the Lower House is recognized both types pool and propose $y^* = X_u + \sqrt{(X_u - Q)^2 - \theta_2^2}$, which both types of the Upper House accept.*

These equilibria are unique. In each of the scenarios, the proposer has a single optimal policy to propose which will be accepted. Other policies might be accepted, but would leave the proposer worse-off.

The proofs and the beliefs which support the equilibria can be found in the appendix. In the following, I discuss the rationale behind Proposition 1.

In an (I,I) scenario, the compromising chamber proposes $y^* = X_u - \theta_j$ when recognized. In this case, the outcome is $x = X_u - \theta_j + \theta_j = X_u$. The compromising chamber is able to realize its own ideal point when recognized.

The accommodating chamber proposes $y^* = 2X_u - Q - \theta_j$ which leads to the outcome $2X_u - Q$. The compromising House defines the room for maneuver in the political system as the policy proposed by the Lower House is the Status Quo mirrored around X_u. It is the point within the winset which is the closest to the ideal point of the accommodating Lower House, hence, the Lower House cannot do better than to propose this policy. This outcome is only possible because both chambers know the true value of θ in a (I,I) scenario.

The policy $y^* = 2x_h - Q - \theta_j$ is not feasible in an (I,NI) scenario. Simply proposing $2X_u - Q + E[\theta]$ instead is also no optimal strategy for the Lower House as for $\theta = \theta_2$ this policy will be vetoed by the House. Remember that the Lower House is the accommodating chamber and can do better if no veto occurs because every movement in the direction of its ideal point leaves the Lower House better-off. Thus, both types pool and propose $y^* = 2X_u - Q - \theta_2$.

The Upper House accepts this proposal. When the Upper House is recognized it proposes $y^* = X_u - \theta_j$. Both types of House in equilibrium are able to get their ideal point when recognized. The Lower House always accepts the proposal.

The Lower House's uncertainty over the *State of the World* creates a strong constraint in combination with the presence of the *informed* House. Comparing the range of feasible policies and the expected outcome for (I,I) and (I,NI) scenarios one can see that both are smaller in the (I,NI) history as the Lower House can no longer surely get its ideal policy implemented. This is a first indication of why interest groups can change the room for maneuver in the political system by strategic communication.

The situation is different if none of the chambers is informed. The best the Upper House can do once it is recognized is to propose its own ideal point minus the expected value of θ. As the latter is zero, the equilibrium proposal is simply X_u.

For the Lower House the matter is more complicated. The Lower House has to propose a policy which makes the Upper House indifferent between accepting the policy and rejecting it. Given the House's risk aversion, this is a point closer to the Upper House's ideal point than would be necessary if the *State of the World* was known to the House. In equilibrium, the Lower House proposes $y^* = X_u + \sqrt{(X_u - Q)^2 - \theta^2}$ which the House accepts.[4] The policy y^* is the Upper Houses' certainty equivalent of the Status Quo.

When only the Lower House is informed, the Upper House will propose its ideal point minus the expected value of θ upon being recognized. The Lower House accepts this proposal. Note that the fact that the Lower House is informed does not help the Upper House in any way.

The Lower House's best response is to propose a policy $y^* = X_u + \sqrt{(X_u - Q)^2 - \theta_2^2}$, which is the same as in the case of two uninformed chambers. The reason for this is that the Lower House has to make the House indifferent between the Status Quo and its proposal. In the case of the Lower House being informed there is no signaling possible as the θ_1 type Lower House has incentives to misrepresent its type.

The uninformed Upper House constrains both the informed and the uninformed Lower House in the same way. The constraint results from the fact that the Upper House determines the room for maneuver in the political system. The Lower House's information level does not constrain the Upper House. As such, the effect of information in the political system is asymmetric.

4.3   ANALYSIS OF THE FULL GAME

The integration of the bargaining stage with the signaling model of lobbying is straightforward. The interest group maximizes its expected utility. This utility depends on the expected outcomes and the costs of sending a message. The costs in turn depend on the strategy chosen.

There is some differences to a standard signaling model of lobbying, such as Potters and van Winden (1992) and Grossman and Helpman (2001). In their setup the interest group is perfectly informed about what will be the outcome of the game, i.e. it knows the policy the decision-maker will

---

[4]In the equation $var(\theta)$ was substituted by θ_2^2.

implement for sure. This is no longer true in my model. The outcome of the game depends on the moves at the bargaining stage. The interest group induces a lottery, where—contingent on its communication strategy and the state of the world—one of two possible outcomes results with equal probability.

With two decision-makers, the rationale of the model is very similar, but there are three cases to consider. Note that the costs may be different for different targets of communication. Let C_u denote the costs associated with a private message directed to the Upper House, let C_l be the costs associated with a private message directed to the Lower House, and let C_p be the costs associated with a public message to both chambers.

The range of lobbying costs is determined by two components. One is the level of fundamental uncertainty. The higher the level of fundamental uncertainty, the higher are the costs caused by the variation in policy outcomes. The other component is the conflict of interest between the decision-maker and the interest group expressed by the distance of their ideal points.

Conditions for Separating Equilibria

Public Communication
The first case to analyze is the interest group's decision to communicate publicly with both chambers. In this case, the interest group does not know the exact policy outcome resulting from the strategy. But it knows the expected policy outcome which results from the strategy. The decision to send a message or not therefore depends on an evaluation of the expected policy outcome (compare also Koessler 2008).

It is important to note that the separating equilibrium depends on the decision-maker taking the interest group at face value. Whenever the interest group sends a message, the decision-makers infer that the State of the World is θ_2 and both implement the respective policies.

The interest group may therefore not have an incentive to lie, that is to send a message when the State of the World is θ_1. This is true when the expected utility from the policy outcome induced by the message minus the message costs is higher than the utility from the policy outcome induced by not sending a message. When $\theta = \theta_2$, the lobbyist truthfully reports the state of the world (i.e. sends a message) if and only if

$$EU_g\big(E[x_2^p|\theta = \theta_2]\big) - C_p \geq EU_g\big(E[x_1^p|\theta = \theta_2]\big) \qquad (4.2)$$

where $E[x_2^p|\theta = \theta_2]$ denotes the expected policy outcome after a public message if the state of the world is θ_2 and $E[x_1^p|\theta = \theta_2]$ denotes the expected policy outcome for the state θ_2 if the chamber implement the equilibrium policies for the state θ_1. This yields the first condition

$$p_1 \equiv -4\theta_2^2 - 4\theta_2 \left(\frac{(X_u - X_g) + (2X_u - Q - X_g)}{2} \right) \leq C_p \qquad (4.3)$$

At the same time, the interest group may not have an incentive to refrain from sending a message whenever the State of the World is θ_2. That is, the expected utility from the policy outcome induced by not lobbying must be higher than the expected utility from the outcome induced by lobbying minus the costs of lobbying. When $\theta = \theta_1$, the lobbyist truthfully reports the state of the world (i.e., sends no message) if and only if

$$EU_g(E[x_1^p|\theta = \theta_1]) \geq EU_g(E[x_2^p|\theta = \theta_1]) - C_p \qquad (4.4)$$

where $E[x_1^p|\theta = \theta_1]$ denotes the expected policy outcome after a public message if the state of the world is θ_1 and the chambers follow their equilibrium strategies. $E[x_2^p|\theta = \theta_1]$ denotes the expected policy outcome for the state θ_1 if the chambers implement the equilibrium policies for the state θ_2. This yields the condition that

$$C_p \leq 4\theta_2^2 + 4\theta_2 \left(\frac{(X_u - X_g) + (2X_u - Q - X_g)}{2} \right) \equiv p_2 \qquad (4.5)$$

Taken together, the incentive compatibility constraints (4.2) and (4.4) determine the range of costs for which a separating equilibrium with public communication exists. Both conditions must hold for a separating equilibrium with public communication.[5] More formally,

$$p_1 \leq C_p \leq p_2 \qquad (4.6)$$

Whenever the costs for lobbying are in the range defined by condition (4.6) a separating equilibrium with both chambers in public exists. In the

---

[5] It is possible to construct an equilibrium, where the chambers do the opposite of what is signaled and the interest group sends a message in State $\theta = \theta_1$. This case is a mirror of the case discussed here and provides no additional insights. Besides, it seems quite unreasonable that lobbying works this way.

case of two decision-makers with random recognition, the costs are evaluated by the distance of the lobbyist to the expected outcome of the bargaining between the chambers.

Private Communication

In the same manner, the conditions for a separating equilibrium with private communication with the Upper House can be derived. If uncertainty is low, the interest group will communicate truthfully in equilibrium if and only if

$$EU_g(E[x_2^u|\theta = \theta_2]) - C_u \geq EU_g(E[x_1^u|\theta = \theta_2]) \tag{4.7}$$

where $E[x_2^u|\theta = \theta_2]$ denotes the expected policy outcome after a private message to the Upper House if the state of the world is θ_2 and $E[x_1^u|\theta = \theta_2]$ denotes the expected policy outcome for the state θ_2 if the chambers follow the equilibrium strategies for the state θ_1 in the state θ_2.

$$EU_g(E[x_2^u|\theta = \theta_1]) - C_u \leq EU_g(E[x_1^u|\theta = \theta_1]) \tag{4.8}$$

where $E[x_1^u|\theta = \theta_1]$ denotes the expected policy outcome after a private message to the Upper House if the state of the world is θ_1 and $E[x_2^u|\theta = \theta_1]$ denotes the expected policy outcome if the chambers follow the equilibrium strategies for the state θ_2 in the state θ_1.

For an equilibrium to exist it must therefore be true that

$$uh_1 \equiv -2\theta_2^2 - 2\theta_2(X_u - X_g) \leq C_u \leq 2\theta_2^2 + 2\theta_2(X_u - X_g) \equiv uh_2 \tag{4.9}$$

The range of costs for which an equilibrium with private communication exists is always smaller than in the case with one decision-maker. The interest group has fewer incentives to send a private message with two decision-makers. This is intuitive given that the interest group anticipates the outcome of the bargaining and the individual decision-maker who is targeted is only partly responsible for the final outcome.

The informational asymmetry between the chambers constrains the Lower House even more compared to the situation where both chambers are informed. The range of feasible policies and thus the expected outcome of the bargaining are smaller in an (I,NI) scenario compared to an (I,I) scenario. The Lower House always proposes the same policy, independently of the State of the World. In consequence, the range of costs for which an equilibrium with private communication with the Upper House exists is smaller than the range for public communication with both chambers.

To complete our discussion of the conditions for separating equilibria we have to consider the possibility for a separating equilibrium with the Lower House in private. For a separating equilibrium with private communication with the Lower House to exist it must be true that

$$EU_g(E[x_2^s|\theta = \theta_2]) - C_l \geq EU_g(E[x_1^u|\theta = \theta_2]) \tag{4.10}$$

where $E[x_2^s|\theta = \theta_2]$ denotes the expected policy outcome after a private message to the Lower House if the state of the world is θ_2 and $E[x_1^u|\theta = \theta_2]$ denotes the expected policy outcome if the chambers follow the equilibrium strategies for the state θ_1 in the state θ_2.

$$EU_g(E[x_2^s|\theta = \theta_1]) - C_l \leq EU_g(E[x_1^s|\theta = \theta_1]) \tag{4.11}$$

where $E[x_1^s|\theta = \theta_1]$ denotes the expected policy outcome after a private message to the Lower House if the state of the world is θ_1 and $E[x_2^s|\theta = \theta_1]$ denotes the expected policy outcome if the chambers follow the equilibrium strategies for the state θ_2 in the state θ_1. This leads to the observation that the following must be true:

$$0 \leq C_l \leq 0 \tag{4.12}$$

Condition (4.12) is different from the other conditions. It states that unless lobbying costs are zero, no separating equilibrium exists where only the Lower House is targeted. In other words, there may be a cheap talk equilibrium, but no signaling equilibrium where the two types separate. Only pooling equilibria exist with the accommodating chamber in private.

The rationale behind this result is the following. The Upper House is the compromising chamber. Its position relative to the Status Quo defines the room for maneuver in the political system (i.e. the size of the winset). Thus, the ignorance of the Upper House constrains *both* the informed and the uninformed Lower House in the *same* way. The best the Lower House can do is to propose a policy that makes the Upper House indifferent between accepting the policy and sticking to the Status Quo. Due to the risk aversion of the Upper House the optimal strategy for the Lower House is to propose the certainty equivalent of the Status Quo. This however, is the same outcome which results if the Lower House proposes when both are uninformed. Paying costs to influence the (accommodating) Lower House necessarily reduces the interest group's payoff compared to a situation where it chooses not to lobby at all. Lemma 1 summarizes this result.

Lemma 1 *Whenever the costs of sending a private message are strictly positive, no separating equilibrium exists in which the interest group privately communicates with the accommodating Lower House.*

Proof See appendix.

In anticipation of the calculus of the interest group, the range for which *public* communication with the Upper House is possible is strictly larger than the range for which *private* communication with the compromising House is possible. In this sense, the presence of the second chamber enables interest group communication compared to the case with one decision-maker. In the terminology of Farrell and Gibbons (1989) we observe one-sided discipline. At the same time, the presence of the first chamber constrains the policy choices of the second chamber. The possibility of communication with the accommodating Lower House is reduced. Farrell and Gibbons (1989) would have coined the term subversion for this kind of result.

The Choice of Lobbying Strategies

So far I have derived the conditions for the existence of separating equilibria. In the benchmark case with one decision-maker, informative lobbying is possible whenever the costs allow for a separating equilibrium with the respective chamber. Here, the condition is necessary and sufficient. The presence of the second chamber constrains the possibility for equilibria. In equilibrium, whenever $p_1 \leq C_p \leq p_2$ a public message results in both chambers being informed about the State of the World as the interest group will only send the message in state θ_2.

However, whenever the interest group sends no public message four possible scenarios can arise. If the cost structure allows for a separating equilibrium with both chambers in public and there is no separating equilibrium with any of the chambers in private, both chambers are informed about the State of the World, as no message is sent if $\theta = \theta_1$.

When the cost structure allows for a separating equilibrium with public communication and for a separating equilibrium with the Upper House with private communication, the Upper House learns the State of the World ($\theta = \theta_1$). The Lower House knows that there is no separating equilibrium with private communication involving itself. So not observing a message may mean that the Interest Group is not sending a message as $\theta = \theta_1$. But it may also imply that the interest group is sending a private message to the Upper House, in which case $\theta = \theta_2$. The Lower House is able to infer the State of the World, when no public message is sent, if and only if the conditions for a separating equilibrium with public communication are met, while the conditions for a separating equilibrium with the Upper House in Private are violated. That is, $uh_2 < C_u$ and $p_1 \leq C_p \leq p_2$.

The rationale behind this argument is that if communication is private, all actors who are not part of the communication situation have no knowledge about the content of the communication. One could imagine a private meeting of the head of a party faction and a lobbyist in a restaurant in Berlin. Other politicians will not know about the existence or the content of the meeting. They will therefore not know whether the interest group engaged in influential communication.

Pooling Equilibria

If there is no separating equilibrium either in private or in public, the result is a pooling equilibrium, where the interest group never sends a message, either private or public. This behavior occurs irrespective of the State of the World. In consequence, none of the chambers becomes informed. The result is a (NI,NI) scenario.

We have now defined the equilibria of the game and hence the choice of lobbying strategies. Proposition 2 summarizes the equilibria.

Proposition 2 *Equilibria of the full game*

(a) *If $|\theta_2| \leq |X_u - Q|$ a separating equilibrium with both chambers in public exists if and only if condition (4.6) holds. The accommodating Lower House will only learn its type in State θ_1 when $C_u > 2\theta_2^2 + 2\theta_2(X_u - X_g)$.*

(b) *A separating equilibrium with the compromising chamber (the Upper House) exists if and only if condition (4.9) holds. In this case, the interest group sends only messages to the House.*

(c) *No separating equilibrium with the accommodating chamber (the Lower House) in private exists.*

(d) *Pooling equilibrium: When none of conditions (4.6, 4.9, 4.12) holds, no separating equilibrium—either in private or in public—exists. The types of interest group pool and do not send any messages in equilibrium.*

Proofs and the corresponding beliefs which support the equilibria can be found in the appendix.

4.4   Implications of the Model

The presence of the second chamber changes the strategic incentives for interest groups compared to a setting with one decision-maker. First, the presence of the second chamber creates *process uncertainty*, defined as uncertainty over the outcome of the political process. It affects interest groups willingness to lobby. The range of costs for which the group is willing to lobby is reduced compared to a model with one decision-maker if the decision-maker were the accommodating actor. In contrast, the range of costs for which interest groups are willing to lobby is increased in comparison to the compromising actor. This finding depends on the interest group's ability to choose the mode of the messages. There is no separating equilibrium in private communication with the accommodating Lower House, hence really progressive interest groups are constrained in their ability to lobby. The result is driven by the fact that the accommodating chamber is unable to signal the *State of the World* to the compromising chamber at the bargaining stage.

The range of costs for which an equilibrium with private communication with the compromising chamber exists is strictly smaller than the range of costs for which a separating equilibrium with public communication exists. This result follows directly from a comparison of Eqs. (4.6) and (4.9). Thus, the possibility to send public messages increases the range of interest groups who are able to send public messages. In particular, there exist separating equilibria with public communication for a range of costs for which no lobbying is possible with either chamber in private. This is equivalent to what Farrell and Gibbons (1989) termed *mutual discipline.*

Second, the presence of a second chamber adds another dimension to the strategic calculus of the interest group. The interest group is able to determine the distribution of information in the system by choosing to communicate privately with one or publicly with both chambers. The choice of the communication strategy is equivalent to choosing among lotteries over policy outcomes. By inducing a distribution of information in the system, the interest group is able to influence the range of possible policy outcomes and in consequence the expected policy outcome. There is, however, not a lot of variation in expected outcomes. More importantly, the interest groups actions reduce the variance of outcomes. The utility gains from the reduced variance in outcomes trump the gains from changes in expected policies due to the risk aversion of the interest group.

This is an important feature of the model and may explain many of the non-findings in empirical research. If the expected outcome is (almost) the same with or without lobbying, a mean centered idea of causality will never lead to strong empirical insights. If one defines a causal effect in terms of the difference in variances as Braumoeller (2006) does, one may be able to find empirical evidence on the effectiveness of interest groups. This approach calls for a comparative approach. The model strongly suggests to direct the effort more into this direction. A strategy for comparative empirical research would be to identify and operationalize the degree of process uncertainty and work from there.

On the other hand, the interest group cannot prevent a change of the Status Quo by not lobbying. Likewise, they are unable to induce bargaining or prevent it from happening, whether they mobilize or not. This is in line with many empirical findings for example (Baumgartner et al. 2009) and offers an alternative explanation for the many non-findings when it comes to study interest group influence.

One question is particularly relevant for my empirical analysis. When do interest groups do prefer private over public signals depending on their ideal position? In order to derive this comparative static one needs to see how the difference in utilities changes if the ideal point moves into a more progressive direction. For a comparison of private messages to the Upper House *vs.* public signals follows

$$EU_g^p(x^*) - C_p \le EU_g^u(x^*) - C_u$$

This condition implies that

$$C_p \le C_u + 2\theta_2^2 - 2\theta_2(2X_u - Q - X_l) \tag{4.13}$$

Differentiating with respect to X_g yields

$$\frac{\partial U_g^p(\cdot) - U_g^h(\cdot)}{\partial X_g} = -2\theta_2 < 0$$

which holds for a given difference in the costs of sending private and public messages. The condition implies that the more progressive an interest group, the more it should prefer private signals over public signals. Similarly, groups opposing a change of the Status Quo are ceteris paribus more likely to engage in public communication. The rationale is that in order to move policy, it is more important to make sure that the compromising

Upper House is informed, as this increases the room for maneuver in the political system.

As there is no separating equilibrium with the accommodating Lower House, the interest group will never prefer a private message to the Lower House over a public message to both.

So far we have analyzed the model by allowing the costs to vary. It is informative to change the perspective and ask, which interest groups (in terms of ideal points) are able to communicate with the decision-makers in public or private for a given cost. This can be achieved by rearranging the equilibrium conditions. It follows that for public messages

$$\bar{X}_g \leq \frac{X_u + 2X_u - Q}{2} - \frac{C_p}{2\theta_2} + 2\theta_2 \qquad (4.14)$$

The separating equilibrium exists only for interest groups with an ideal point to the left of $\bar{X}_g^p$. Note that the threshold is a function of the expected policy outcome $\frac{X_u + 2X_u - Q}{2}$ as well as the level of uncertainty and the costs.

A separating equilibrium with private messages to the compromising Upper House exists for interest groups with an ideal point smaller than

$$\bar{X}_g^u \leq X_u + \theta_2 - \frac{C_u}{2\theta_2} \qquad (4.15)$$

for a given cost C_u. In this case, $\bar{X}_g^u$ is a function of the ideal point of the compromising actor, as well as the costs and the level of uncertainty.

These results will be important in the empirical analysis as well.

4.5   HYPOTHESES

So far I have discussed the theoretical model in abstract terms. In order to confront the theoretical predictions with reality, it is necessary to derive hypotheses which are open to empirical investigation.

One of the main empirical challenges in interest group research is to make sense of the empirical patterns of interest group communication activities. Once coded in a dataset one has to deal with an aggregate of individual decisions. There is considerable variation in both mobilization and activity patterns across legislative events in Germany. In order to construe patterns in the data, it is important to understand the underlying problem structure and the nature of those shifts in lobbying patterns.

McFadden (1974, 106) argues that if we observe systematic variation in aggregate choice with discrete alternatives, this must stem from shifts in individual choice at the *extensive* margin, resulting from a *distribution* of decision rules in the population. Applied to lobbying, this means that, whenever there is a systematic variation in lobbying activities across policy events, then this variation must stem from individual choices determined at the extensive margin. The variables which influence choices at the extensive margin have been discussed thoroughly. The most important aspect is the *decision rule*. It entails two interrelated decisions. One is to mobilize and the second is the decision to send either a private or a public message. I have set up a theoretical model to describe this decision rule.

While the decisions are not fully separable in practice I have shown that one can fruitfully do so for analytical purposes in order to derive predictions on the chain of decisions leading from MOBILIZATION to ACTION. The two parts of an interest groups strategy follow two distinct utility calculations:

Mobilization This part is similar to the logic of any signaling game. The necessary condition is that the utility from acting is higher than the utility from not acting ($U(Action) > U(No\ Action)$). This implies that given the state of the world, the interest group has an action at its disposal which it can carry out and is strictly better-off compared to not carrying it out. There may, however be different courses of action.

Action I identify two classes of interest group actions: Public and private messages. Both may encompass different activities which are structurally equivalent. The groups choose a public message if and only if $U(public) \geq U(private)$ and $U(Action) > U(No\ Action)$. Likewise, they choose a private message if and only if $U(private) > U(public)$ and $U(Action) > U(No\ Action)$. Together, the two inequalities are sufficient to determine the choice of an interest group.

In the following concrete hypotheses are derived. They will be operationalized in the next chapter.

Hypothesis 1 *Interest groups with ideal points to the left of the expected policy outcome are, ceteris paribus, more likely to send public messages. The larger the distance, the higher is the likelihood of a public message.*

Hypothesis 1 follows directly from the equilibrium predictions of the model. The choice of scale in the theoretical model is irrelevant here, as one could invert the scale and yield the same directional results. The distance measure is defined as in the model. That is, a positive distance indicates an ideal point to the left of the expected policy outcome.

Note that 'to the left of' is understood in spatial terms, not in ideological terms, although in my application both terms coincide. The expected effect of the distance variable is *negative*.

The equilibrium prediction of the theoretical model would be even stronger: It would predict that interest groups with ideal points to the right of the expected policy outcome would not send messages at all. Empirically, this is too strong given the measurement error and the impossibility to exactly locate the Status Quo.

I also do not explicitly model interest group competition which may affect the equilibrium outcomes. The fundamental underlying mechanism can nevertheless be expected to be traceable in the data.

Hypothesis 2 *Interest groups with ideal points to the left of the ideal point of the compromising actor are, ceteris paribus, more likely to send private messages compared to interest groups with ideal points larger than the compromising actor. The larger the distance, the higher the likelihood of a private message.*

The underlying mechanism is the same as in the case of a public message. However, as explicated in the theory, the model predicts that the compromising actor is the constraining actor. Privately lobbying the accommodating actor does not change equilibrium outcomes. Here too, the theoretical model would predict a sharp drop in activity at the position of the compromising actor.

In its extremity also this prediction is surely too strong in the empirical world. Note also, that for very low costs (close to zero) interest groups might be willing to lobby the accommodating actor.

Hypothesis 3 *The more an interest group is in favor of a policy change (i.e. the more the group's ideal point is to the right), the more it, ceteris paribus, prefers private over public messages.*

Interest groups may choose to send private or public messages in equilibrium, as long as the conditions for a separating equilibrium hold. I have shown that more progressive interest groups should ceteris paribus prefer private over public messages. This would contradict hypothesis two, which is based on the condition for the existence of the separating equilibrium.

Taking the overall strategic incentives into account changes the prediction. It will be particularly interesting to test this difference.

Hypothesis 4 *The higher the lobbying costs, the lower, ceteris paribus, the probability that an interest groups sends either public or private messages.*

In signaling games, the costs are a central element of information transmission. Taking higher costs potentially strengthens the signal. At the same time, higher costs reduce the willingness of a specific type to engage in costly lobbying as the gains from signaling the state of the world are reduced.

In the lobbying model, the gains are fixed and are determined by the interest group's distance to the policy outcome. Higher costs should therefore reduce the willingness of interest groups to lobby ceteris paribus.

This is equally true for both public and private messages. There is no way to separate the costs for private and public messages empirically. I therefore use the same cost variables for both types of messages.

Hypothesis 5 *Interest groups with relatively many access points in the Bundesrat are, ceteris paribus, more likely to send private messages.*

The model predicts that only addressing the accommodating actor is pointless if the interest group tries to influence policies and should only occur if the costs of doing so are zero. The question of access should be irrelevant for public messages. For private messages, better access means to have better access to the compromising actor in all bill that require consent.

Hypothesis 6 *Higher fundamental uncertainty increases the likelihood that an interest group sends a message (either private or public).*

Higher fundamental uncertainty implies that the shock which maps the policy into the outcome is larger. By sending a message the interest group thus has a relatively larger impact on the variation of the policy outcome. This increases ceteris paribus the incentives to lobby as interest groups suffer from higher variation of policy outcomes due to their risk aversion. Likewise, higher uncertainty increases the range of ideal points for which interest groups can credibly signal the state of the world. This follows directly from Eqs. (4.14) and (4.15).

Appendix

Proof of Proposition 1

Proof of Proposition 1a. Upon recognition, the informed Upper House chooses a policy $y \in \text{argmax } U_l(y, \theta)$. The unique optimum is to choose the ideal policy by proposing $y = X_u - \theta_j, j \in 1, 2$ in accordance with the State of the World. The resulting outcome is X_u. As the Lower House prefers the Upper House's ideal point to the Status Quo, it accepts the proposal. If the Lower House is recognized, it proposes $y = 2X_u - Q - \theta_j, j \in 1, 2$ in accordance with the State of the World. This is the policy which maximizes the Lower House's utility conditional on approval by the Upper House, i.e. $y \in \text{argmax } U_l(y, \theta)$ s.t. $x \in \{x | U_u(x, \theta) \geq U_u(Q, \theta)\}$. The Upper House accepts this proposal as it is indifferent between the Status Quo and the resulting policy. □

Proof of Proposition 1b. Upon recognition of informed Upper House, the legislative stage has a separating equilibrium, if $U_u(y_1 | \theta = \theta_1) > U_u(y_2 | \theta = \theta_2)$ and $U_u(y_2 | \theta = \theta_2) > U_u(y_1 | \theta = \theta_2)$, where $y_1 = X_u - \theta_1$ and $y_2 = X_u - \theta_2$. Both conditions hold, as $\text{argmax } U(x | \theta = \theta_1) = \{X_u - \theta_1\}$ and $\text{argmax } U(x | \theta = \theta_2) = \{X_u - \theta_2\}$. The House thus has no incentive to misrepresent the State of the World. For the Lower House it holds that $U(y_1 | \theta = \theta_1) = U(y_2 | \theta = \theta_2) > U(Q, \theta)$.

The Upper house will propose $X_u - \theta_1$ in state θ_1 and $X_u - \theta_2$ in state θ_2. The posterior belief of the uninformed Lower House after a proposal by the Upper House is

$$\mu_l(\theta = \theta_2) = \begin{cases} 1 & \text{if } y = X_u - \theta_2 \\ 0 & \text{if } y = X_u - \theta_1 \\ 0.5 & \text{otherwise} \end{cases}$$

The Lower House thus accepts the proposals in both states of the world as the outcome $x^* = X_u$ is strictly better for the Lower House than the Status Quo.

If the uninformed Lower House is recognized, no separating equilibrium is possible. The Lower House thus chooses the policy which maximizes expected utility. The Lower House is strictly better-off with any policy change compared to the Status Quo. Thus, the Lower House wants to avoid a veto. The problem is therefore to choose the proposal which yields the highest expected utility conditional on not being vetoed. The

House would veto any proposal leading to a policy outcome larger than $2X_u - Q$ as $U_u(Q) = U_u(2X_u - Q)$. The maximum policy the Lower House can propose is thus $2X_u - Q - \theta_2$. The Upper House accepts this proposal as. □

Proof of Proposition 1c. If the House is recognized, it proposes X_u. This strategy is maximizing its expected utility as $E[\theta] = 0$ and $|X_u - Q| > \theta$. Thus, in any State of the World, $Q < X_u + \theta_j < 2X_u - Q$. The House is therefore strictly better-off compared to the Status Quo. The Lower House accepts any of those proposals, as it is also strictly better-off compared to the Status Quo. See the proof of Lemma 1 for a proof of the Lower Houses behavior when facing an uninformed House under low uncertainty. □

Proof of Proposition 1d. See proof of Proposition 1c for the Upper House, and the proof of Lemma 1 for the Lower House. □

Proof of Lemma 1. The uninformed Upper House will accept any proposal, for which it is at least indifferent between the Status Quo and the outcome, i.e. $U_u(Q) \leq EU_u(p)$. Define the set A as the subset of $\mathbb{R}$ for which this is true. A is then given by the interval $[Q + \theta_2, X_u + \sqrt{(X_u - Q)^2 - \theta_2^2}]$. The point that is closest to the ideal position of the Lower House is the upper bound of the interval which is the House's certainty equivalent of the Status Quo. The Lower House is unable to propose a policy which reveals the *State of the World* due to strong incentives to misrepresent the State of the World. To see this assume that the Lower House could signal the State of the world to the Upper House. Define P_1 as $2X_u - Q - \theta_1$ and P_2 as $2X_u - Q - \theta_2$. For a separating equilibrium in the legislative stage it must hold that: (1) $U_l(P_1|\theta = \theta_1) > U_l(P_2|\theta = \theta_2)$, and (2) $U_l(P_2|\theta = \theta_2) > U_l(P_2|\theta = \theta_1)$. The second equation holds, as $-(2X_u - Q + 2\theta_2 - X_l)^2 > -(2X_u - Q + 2\theta_1 - X_l)^2$ whenever $0 < \theta_2$ which is true by definition of θ_2. The first condition does not hold as $-(2X_u - Q - X_l)^2 > -(2X_u - Q + 2\theta_1 - X_l)^2$ is only true if $0 < 2\theta_1$. As by definition $\theta_1 < 0$, this is impossible. The types of Lower House thus pool and propose the policy that is closest to their ideal point, conditional on being an element of A. This is exactly the certainty equivalent of the Status Quo. □

Proof of Proposition 2

Proof of Proposition 2a. The necessary condition (4.6) for a separating equilibrium has to be fulfilled. If this is the case, the following beliefs of the chambers support the separating equilibrium with public communication:

For the Upper House:

$$\mu_u(\theta = \theta_2) = \begin{cases} 1 & \text{if } m \text{ and } p_1 \le C_p \le p_2 \\ 0 & \text{if } n \text{ and } p_1 \le C_p \le p_2 \\ 0.5 & \text{otherwise} \end{cases}$$

and the Lower House:

$$\mu_l(\theta = \theta_2) = \begin{cases} 1 & \text{if } m \text{ and } p_1 \le C_p \le p_2 \\ 0 & \text{if } n \text{ and } p_1 \le C_p \le p_2 \text{ and } (uh_1 \ge C_u \text{ or } C_u \ge uh_2) \\ 0.5 & \text{otherwise} \end{cases}$$

Given these beliefs, and the proposal and veto activities outlined in Proposition 1 the best response of the interest group is to send a public message lobbying whenever the state is θ_2.   □

Proof of Proposition 2b. The necessary condition (4.9) for a separating equilibrium with private communication with the House has to be fulfilled. Once these are fulfilled, sending a private message to the Upper House when the state is θ_2 is a best response to the veto and proposal strategies outlined in Proposition 1 and the beliefs of the Upper House

$$\mu_u(\theta = \theta_2) = \begin{cases} 1 & \text{if } m \text{ and } uh_1 \le C_p \le uh_2 \\ 0 & \text{if } n \text{ and } uh_1 \le C_p \le uh_2 \\ 0.5 & \text{otherwise} \end{cases}$$

and the Lower House

$$\mu_l(\theta = \theta_2) = 0.5$$

The Houses are best responding to the messages of the interest group and each others veto and bargaining strategies.   □

Proof of Proposition 2c. The necessary condition (4.12) for a separating equilibrium has to be fulfilled. As the costs of sending a private message to the Lower House are strictly larger than zero, the necessary condition (4.12) is violated. Therefore, no separating equilibrium with private communication with the Lower House exists.   □

Proof of Proposition 2d. Follows directly from the violation of the incentive compatibility constraints for the separating equilibria.   □

REFERENCES

Baron, D. P., & Ferejohn, J. A. (1989). Bargaining in Legislatures. *American Political Science Review, 83*(4), 1181–1206.

Baumgartner, F. R., Berry, J. M., Hojnacki, M., Kimball, D. C., & Lech, B. L. (2009). *Lobbying and Policy Change: Who Wins, Who Loses, and Why.* Chicago: University of Chicago Press.

Braumoeller, B. F. (2006). Explaining Variance: Or Stuck in a Moment We Can't Get Out Of. *Political Analysis, 14*(1), 268–290.

Esterling, K. M. (2004). *The Political Economy of Expertise: Information and Efficiency in American National Politics.* Ann Arbor: The University of Michigan Press.

Farrell, J., & Gibbons, R. (1989). Cheap Talk with Two Audiences. *The American Economic Review, 79*(5), 1214–1223.

Gilligan, T. W., & Krehbiel, K. (1987). Collective Decisionmaking and Standing Committees: An Informational Rationale for Restrictive Amendment Procedures. *Journal of Law and Economic Organization, 3*(2), 287–335.

Grossman, G. M., & Helpman, E. (2001). *Special Interest Politics.* Cambridge, MA: The MIT Press.

Koessler, F. (2008). Lobbying with Two Audiences: Public vs Private Certification. *Mathematical Social Sciences, 55*(3), 305–314.

Kreps, D. M. (1990). *A Course in Microeconomic Theory.* Herfortshire: Harvester Wheatsheaf.

Matthews, S. A. (1989). Veto Threats: Rhetoric in a Bargaining Game. *The Quarterly Journal of Economics, 104*(2), 347–369.

McFadden, D. (1974). Conditional Logit Analysis of Qualitative Choice Behavior. In P. Zarembka (Ed.), *Frontiers in Econometrics* (pp. 105–142). New York: Academic Press.

Morton, A. (1993). Mathematical Models: Questions of Trustworthiness. *The British Journal for the Philosophy of Science, 44*(4), 659–674.

Potters, J., & van Winden, F. (1992). Lobbying and Asymmetric Information. *Public Choice, 74*(3), 269–292.

Rogers, J. R. (1998). Bicameral Sequence: Theory and State Legislative Evidence. *American Journal of Political Science, 42*(2), 1025–1060.

Romer, T., & Rosenthal, H. (1978). Political Resource Allocation, Controlled Agendas, and the Status Quo. *Public Choice, 33*(4), 27–43.

Tsebelis, G. (2002). *Veto Players: How Political Institutions Work.* Princeton: Princeton University Press.

Tsebelis, G., & Money, J. (1997). *Bicameralism.* Cambridge: Cambridge University Press.

CHAPTER 5

Data and Operationalization

Abstract In this chapter, I discuss the data and the main variables. These include the measurement of preferences of interest groups and political decision-makers. Based on those, distance measures are developed. Lobbying costs and several control variables are operationalized.

Keywords Ideal point estimation · Spatial distance · Centrality Salience · Lobbying costs · Lobbying strategies

5.1 THE DATA

5.1.1 *The Policy Domain: Actors and Interests*

It is now time to confront the theoretical predictions with data. The requirements for a dataset which allows to test the theory are high. It is necessary to contain information on preferences of decision-makers and interest groups. Information on interest group activities on specific policies is likewise necessary. The model explicitly highlights interest group politics in a bicameral setting in a electoral system with high party discipline, therefore all datasets on the US case are not applicable. The best dataset available for testing the theory is the data used by Knoke et al. (1996) for the comparison of lobbying networks Germany, Japan, and the USA. The data were collected by two teams of researchers in Germany and the United States. The main focus was on policy networks in the policy domains of labor and social policy. As my model is only applicable in the German case, I will restrict my analysis to the subset of data on Germany.

© The Author(s) 2019
S. Koehler, *Lobbying, Political Uncertainty and Policy Outcomes*,
https://doi.org/10.1007/978-3-319-97055-4_5

The German part of the project was conducted by Franz Urban Pappi, Thomas König and colleagues at the University of Kiel, Germany.[1] The researchers collaborated closely with David Knoke (University of Michigan, Ann Arbor, USA) and colleagues[2] on the development of the research design and the collection of data.

The data were used in a variety of studies.[3] Most of the publications based on the dataset explicitly analyze the structure or explain the formation of policy networks. Some also try to get to the question of influence within the network structures (e.g. Pappi et al. 1995). Little use was made of the information on interest group activities in previous studies.

The study design and the data collection closely resembled the approach taken by Laumann and Knoke in their study on policy networks in the policy domains of Health and Energy in the United States (Laumann and Knoke 1987). A *policy domain* is defined as consisting of the actors concerned with a substantive policy area (e.g. social policy). A defining characteristic is that the actors' preferences and actions matter in the sense that their views and activities have to be taken into consideration by rest of the members of the policy domain (Laumann and Knoke 1987). A policy domain has a strong issue context, it is defined on shared interests. Interests therefore formed the basis for the identification of the relevant groups to interview for the data collection.

A newspaper research was conducted to identify major issues in the policy domains of labor and social policy in Germany in the 1980s. Then the major actors were identified. Major actors were considered to be all political actors who are directly relevant for decision-making (e.g. party groups in Parliament). Interest groups and organizations who either participated in a hearing at the committee level or were mentioned in newspapers as taking action on the issue were likewise deemed members of the policy domain. All relevant domain actors were then surveyed by the two teams of researchers (Knoke et al. 1996, 66). The total number of surveyed actors in Germany is 126. The reliability of the selection mechanism was cross-checked by

---

[1] For more information see the codebook (König and Pappi 1989).

[2] For more information see the documentation and data at the ICPSR archive: Knoke and J. Kaufmann (1992).

[3] The US labor policy domain was studied in Knoke and Burleigh (1989), Knoke (1990), Knoke and Pappi (1991). The German labor policy domain was studied in König (1992), König and Bräuninger (1998), Bräuninger and König (2004) and for two comparative studies (Knoke et al. 1996; Pappi et al. 1995). See also Pappi and Henning (1998).

eliciting respondents perception of the importance of actors in the policy domain.

This procedure is an elegant way to deal with the problem of identifying the universe of interest groups. Several approaches exist but are inherently problematic (Coen 2007). The restriction to a clearly defined domain and a restriction to the set of interest groups which is considered important by the other actors ensures that the data do not suffer from selection bias.

Response rates in the study of lobbying are usually low. Often they are not even reported, which questions the reliability of the data. Mahoney (2007), for example, does not give any information about how many lobbyists were asked for an interview to achieve the random sample of 149 interviewees. Whether the data represent a true random sample is therefore largely unclear. A response rate of more than 95% for the German groups is an additional factor which underlines the high reliability of the data collected by König and Pappi.

The interest structure of the domain actors was assessed using a hierarchical definition of interests. The three sets of interests are the so-called *subfield interests*, the *issue interests* and interests in so-called *policy events*, i.e. specific legislative processes.[4]

A subfield is a major topic within the policy domain. Examples for the policy domain of labor and social policy would be social policies or labor market policies (Knoke et al. 1996, 81f.). A policy dimension is a (one-dimensional) ordered set of possible policy outcomes (cf. Shepsle and Bonchek 1997, 91f.). The issue structure of a policy domain potentially contains many policy dimensions on which the issue structure is defined. Pappi et al. (1995) have shown that using the subfield interests it is possible to construct two policy dimensions from the data. The first one is the *labor policy* dimension, the second one is the *social policy* dimension. A general distinction between the policy dimensions of labor and social policy is that social policies tend to be (re)distributive while labor policies are mainly regulatory (Lowi 1972, 300).

The second type of interests were the *issue interests*. A policy issue is defined narrower than a subfield, but broader than any given bill within the domain. An example for an issue in the labor policy domain is the

---

[4]For reasons of control some of the events were not legislative processes but court proceedings at the federal court with relevance to the policy domain of labor. In my analyses, I will drop those cases as they are not directly relevant for testing theoretical arguments based on my model.

discrimination of women at the workplace. There are many conceivable courses of action on the issue. Hence, many potential bills to deal with this issue are imaginable. Issues are usually persistent. While a law may be enacted that regulates some aspect of an issue, the issue itself potentially remains salient as not all interested parties agree on the course of action.

For the determination of issue interests, a set of major issues in labor and social policy during the 1980s was identified and the organizations were asked about their position on those issues. The reported issue interests were used to analyze the homogeneity of the policy domain by running a principal components analysis on those interests. In both the US and the German policy domain the first component accounts for the major part of the variance, which indicates a homogeneous policy field.[5]

The issue interests were used to calculate the distance between actors in the policy space using a multidimensional scaling procedure (ALSCAL). This allows to map the spatial distribution of actors in the policy space (see Knoke et al. 1996). Moreover, as Pappi et al. (1995) have shown, the structure of the issue interests is determined by the structure of the subfield interests. As such the sets of interests are coherent.

The third set of interests is the most relevant to my analysis. A major focus of the data collection was the gathering of data on actors' preferences regarding specific policy *events*. A policy event occurs when a bill is proposed and introduced into the decision-making process (Laumann and Knoke 1987). An event thus denotes a critical point in the political process related to a specific bill such as the 'Haushaltsbegleitgesetz of 1984' in Germany. It could be the introduction of the bill per se, a parliamentary hearing, or any other formal activity which is necessary to pass a law.

Knoke et al. (1996) conceptionally distinguish *interests* and *preferences*. They use *preference* as an expression of being in favor of or against a policy. *Interest* denotes the intensity of utility differences stemming from different policies. What they call interest is similar to a salience in a spatial utility function (König and Bräuninger 1998).

Pappi et al. (1995) have used the actors' interests and preferences regarding events to calculate concrete policy positions. The organizations had expressed their interests in the events on a scale from 0 to 5. Zero indicates no interest at all, and five indicates a very strong interest in the event.

---

[5]Note that this is based on the common perception of the actors, not some external criteria. A point also raised by Pappi et al. (1995, 202).

Additionally, actors were asked to state their preferences regarding the policy. It was coded 1 if they were in favor and −1 if they were against the policy. Pappi et al. multiplied the two numbers to derive an index of directed interest which ranges from −5 to 5. This index was then used to identify the policy positions of the actors using a principal components analysis. In my study, I will use a different approach to calculate policy positions of the political actors. I also propose a different use for the *interest* variable.

5.1.2   Types of Data

The procedure to identify the domain actors was driven by actors' interest in predefined issues in labor and social policy and their relevance given by (a) participating in a public hearing, (b) being mentioned in newspaper reports, or (c) being mentioned as important by other relevant actors.

The datasets which result from surveying the domain actors contain two distinct *types* of data. Firstly, in addition to the interests, researchers collected data on actors' characteristics, organizational structure, and resources. Among others, the organizations were asked about their efforts, budget, internal organization and member structure. With respect to the policy events the researchers gathered data on the organizations' actions and goal attainment.

Secondly, the network structure of the policy domain was assessed. The researchers gathered data on two types of network data. Actor networks were collected using contact reports for a couple of relevant aspects (information exchange, resource exchange, influence reputation). One important aspect is that respondents were asked about both incoming and outgoing contacts with other organizations. These networks allow to map the general relationship between actors in the policy domain.

Other networks which were surveyed are bipartite networks of event publics (attendance in events). Systematic data on event-specific coalitions were also collected. An *event public* consists of all domain organizations that express interest in a particular policy event, regardless of which outcomes they prefer (Knoke and Pappi 1991). The data allow to analyze event-specific mobilization and activity patterns. The data also contain information on event activities and mobilization. They have surprisingly been underutilized.

In addition, I coded some variables myself. The variables describe the political process in more detail. For example, whether the bill was subject to approval or not, the number of committees involved among others.

Despite their age, the data are the best data available to test my theoretical claims. Some people may argue that technological change has increased the possibilities of interest groups to engage in communication activities. The internet and modern telecommunications may have radically changed lobbying tactics, which may now involve mass emails or online petitions.

The potential criticisms do not bite in my case. Firstly, my model is general. It distinguishes activities on the basis whether they are public or private, i.e. the mode of the message. From this point of view, sending a letter or an email are *functionally equivalent*. Likewise, mobilizing constituents based on a mailing to group members or using a web page are functionally equivalent. For example, google opposed the introduction of a bill on the so-called *Leistungsschutzrecht* in late 2012. The bill was supposed to regulate copyright issues related to the display of excerpts of protected works in search results. Google tried to mobilize customers with the words 'defend your internet!' and linked campaign materials on their search page. However, providing information with a mailing to one's members and asking them to take action is no different, just much more expensive. Technological changes do not invalidate the empirical approach as they are mostly affecting the *costs* of communication.

Secondly, not many datasets on interest group activities are available. Of those datasets available, most lack either information on the strategies used by interest groups (e.g. Mahoney 2007) or they lack information on preferences of interest groups regarding specific events (e.g. Dür and Mateo 2013). Both are vital for testing predictions of the model. As such, the dataset of König and Pappi (1989) is quite unique and the most appropriate to confront the model with empirical data.

5.2   Dependent Variables

5.2.1   Interest Group Activities

Interest group communication strategies can comprise several activities. The dataset contains six activities which describe interest groups' behavior. All six serve as dependent variables in my analysis. The questions which were asked to code the variables are summarized below[6]

---

[6]Compare the German questionnaire (König and Pappi 1989, 29–32).

Content Did the organization provide content/participate in the formulation of the bill?

Informal Did the organization have informal contacts with government officials or parliament?

Formal Did the organization have formal contacts with government officials or Parliament (e.g. giving testimony at hearings/serving on commissions)?

Massmedia Did the organization use mass media to announce its opinion about the event?

Mobilization Did the organization mobilize members or the general public?

Coalition Did the organization form coalitions with other groups?

The first dependent variable codes whether the actor/group had been involved in the formulation of the bill. This is clearly of major interest to my model as it is the core of informational lobbying. Interest groups provide inputs to a bill because they have specialized knowledge and specific expertise which decision-makers lack. By the nature of law, the provision of content has to be precise. Therefore, the content is usually provided in written form. This is only observable if it is made public by the receiver of the documents which is unusual. I therefore treat the provision of content as sending a *private* message.

Informal contacts with decision-makers are the second dependent variable. The major purpose of informal contacts is informational. The lobbyist uses the contact to communicate his position on the issue and likely consequences of the proposed bill. This analysis provides insights into determinants of direct influence attempts. Unfortunately, the wording does not allow to fully discriminate between different targets. I therefore use additional arguments to derive hypotheses about interest group behavior regarding private messages.

The third dependent variable, formal contacts, codes whether interest groups use formal channels of influence attempts to communicate information. This includes participation in commissions or public hearings. As these activities are publicly observable, they could constitute a *public* message. There is, for example, a target problem. When testifying in a hearing it is generally impossible to specifically target only one of the party factions in the committee (that does not mean that a statement may affect different actors differently). Secondly, as formal acts are usually either subject to media observation or subject to official archiving, e.g. in protocols, reputation effects will be more important in those cases. However, there is an additional step involved: Formal activities have a strong demand component. After all, one needs to be invited to testify in a committee hearing. The activity is therefore somewhat different. I therefore do not characterize it as part of a *public* message.

The use of massmedia has two relevant aspects. Firstly, a strong mobilization effect. Media usage might influence the salience of a topic or the position of voters. An example for this is the campaign for introducing a minimum wage in Germany that was started by the German Trade Union Confederation (DGB) in 2007. The aim of this campaign is not so much to educate policymakers but to put the topic on the political agenda and to mobilize voters. Secondly, media usage is useful to communicate positions publicly. This publicity allows to target legislators the interest groups do not have direct access to. It also increases the reputation costs for the interest groups. A false statement made in public may have dramatic effects on the trustworthiness of interest groups.

The mobilization of members can have two aspects. One is to exert pressure on politicians by asking members to give a supportive or critical opinion on the current bill and communicate this information to their representative or to a minister. This strategy is often used by organizations like Greenpeace or Amnesty International, who try to strengthen the legitimacy or weight of their position in this way. The main message is encoded in the willingness of the interest group to bear the costs for the mobilization. The costs of mobilization do not only encompass the material costs for making calls, sending letters or the like. A major aspect of the costs is that it is impossible for groups to use mobilization strategies on a regular basis without threatening the willingness of their constituency to participate. Therefore, there is a long-term cost of a reduced usefulness of the strategy in the future. The mobilization (or at least the results of the mobilization strategy) is usually publicly observable, so mobilization can be characterized as *public*.

The last variable is coalition building. Joining a coalition can also be interpreted as a communicative act (Esterling 2004). Firstly, it signals that more than one relevant actor shares certain concerns and not only a radical minority. Coalitions thereby provide important information to policymakers about the general support of the bill at stake. Secondly, it also signals that interest groups are willing to overcome collective action problems. The costs (and thus the information content) are even stronger in case of an unlikely coalition, e.g. a coalition of trade unions and employers' asso-

Table 5.1 Aggregation of dependent variables

Public	Private
massmedia	content
mobilization	informal
coalition	

ciations. Coalitions are usually public knowledge, as interest groups sign common position papers, give common press conferences or give reference to the mutual support. Coalition building can therefore be seen as a *public* message.

5.2.2   Aggregation

While I provide regression results for the individual activities, I am not interested in the individual activities per se, but in the question whether interest group sends a public or a private message. The model is concerned with costly actions, not with specific communication acts.

I aggregate the activities into an index for *public* and *private* activities (see Table 5.1). The variable public was created by aggregating the variables massmedia, mobilization, and coalition. The variable was coded as one if at least one of the actions had been carried out by the group and zero if none of these actions had been carried out. The variable private was coded correspondingly by aggregating the variables content and informal.

There are two approaches to the aggregation of the information. The first approach is to use a binary variable which is one if at least one of the actions in the respective category was chosen. I did this for both the private and public messages. Whenever I refer to a public or private message in the empirical chapter I refer to the indicators derived in this way. With this operationalization, all information about the strength (or costs) of the intervention is lost. In order to check the severity of the problem, two additional indicators are developed.

The first indicator is a simple sum. For the public message this indicator ranges from zero to three. For the private message it ranges from zero to two. This indicator also accounts for the strength of the reaction, treating all activities as equal. This approach, however, may not be unproblematic as well. The messages may themselves differ in costs.

The data are structurally similar to educational testing data. To measure the relative costs of the tactics an item response theory (IRT) model is used. The IRT model is the appropriate measurement model for this type of data (Jackman 2008; Johnson and Albert 1999).

The model of choice is a Rasch model which is a one parameter specification of the IRT model. It allows to estimate a difficulty parameter for each activity. The difficulty parameter measures how likely it is that a correct response is given, i.e. how likely an interest group is to use the strategy. Higher difficulty parameters imply a lower likelihood for a given level of resources and ability of the interest group. The difficulty parameter can be interpreted as the costs of using the respective tactic. An item with a higher difficulty parameter is more costly.

Table 5.2 shows the results of the Rasch model. The difficulty parameters suggest that the first three variables (content, informal, formal) are roughly comparable in difficulty (and therefore costs) massmedia is a middle category mobilization and coalition are the hardest tasks.

The corresponding item characteristic curves resulting from the model are shown in Fig. 5.1. The curves for content and informal are so close to each other that the latter curve is overlaying the former in the plot. The Rasch model rests on the assumption that all items discriminate equally well between groups (Johnson and Albert 1999). The more a curve is to the right, the lower the likelihood that the strategy is used at each specific ability level. This suggests that the public strategies are harder, i.e. more expensive than the private strategies. Public strategies are used frequently, which is in line with the prediction that separating equilibria with public communication exist for a wider range of costs than equilibria with private communication.

The results of the Rasch model are used to construct a weighted version of the strategies. As the private activities are equally costly, the weighted version is the same as the count version. For the public strategies, the weighted version is created by multiplying the more costly strategies (coalition and mobilization) by two, before summing up. The result is a variable ranging from one to five.

Figures 5.2, 5.3 and 5.4 show the joint distribution of public and private strategies for the three operationalizations. The mosaic plot in Fig. 5.2 shows that a majority of interest groups engages in both types of activities. The second largest share is interest groups who are only sending private messages. The number of groups sending none or only public messages are relatively small. The case in which both types of messages are sent

Item Characteristic Curves

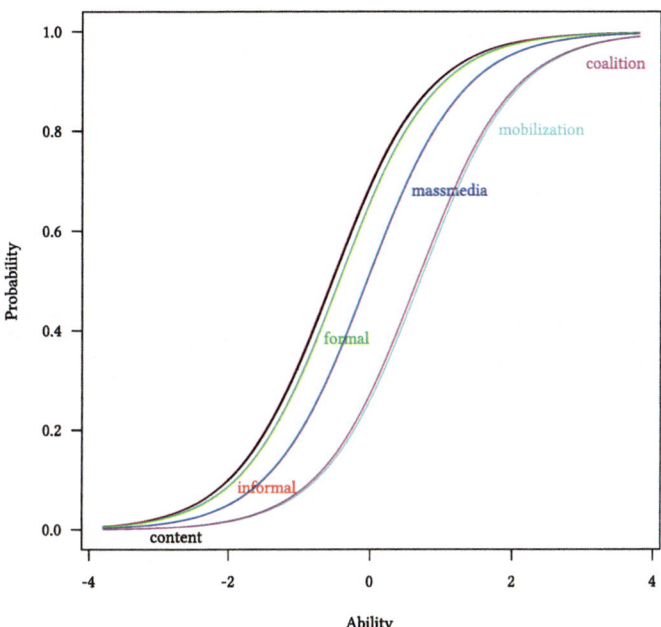

Fig. 5.1 Item characteristic curves of dependent variables resulting from a Rasch model

Fig. 5.2 Joint distribution of dependent variables (binary version)

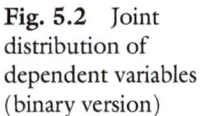
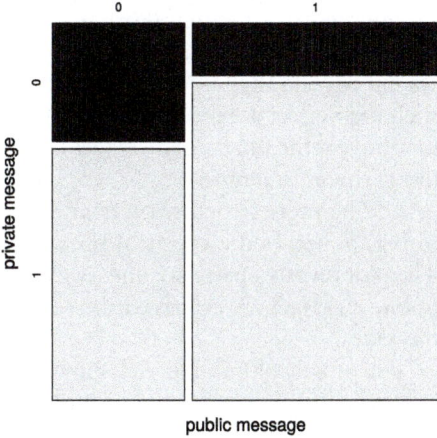

Table 5.2   Results of item response analysis for the lobbying tactics (Rasch model)

Variable	Difficulty parameter	Std. error	z-value
content	−0.5565	0.0601	−9.2567
informal	−0.5601	0.0602	−9.3101
formal	−0.4343	0.0587	−7.3973
massmedia	−0.0723	0.0565	−1.2800
mobilization	0.6618	0.0613	10.7901
coalition	0.6321	0.0609	10.3786

Fig. 5.3   Joint distribution of dependent variables (count version)

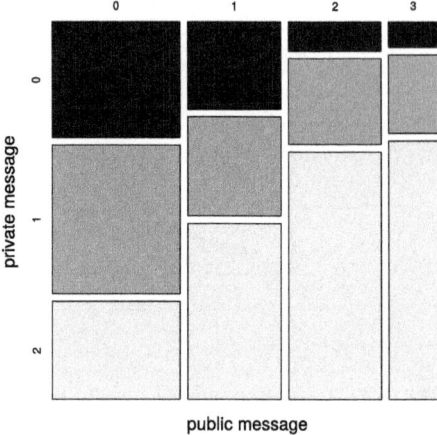

public message

is equivalent to the sending of a public message. Public messages trump private messages in the sense that they increase the information in the system beyond the information content created by the private message. As such their effect is expected to be stronger. One reason for this observation may be that the timing of messages is not simultaneous, a point not reflected in my theoretical model.

Adding more structure on the definition of the strategy, the mosaic plot in Fig. 5.3 gives the count of the activities an interest group engaged in. The plot reveals a pattern quite similar to the binary world. More expensive public messages are ceteris paribus accompanied by more expensive private messages.

The mosaic plot in Fig. 5.4 shows the joint distribution of the weighted strategies. The plot shows that there is somewhat more variation as to the relative weights of the strategies. Some combinations are more common than others. The pattern is also increasing. More expensive private mes-

Fig. 5.4 Joint distribution of dependent variables (weighted version)

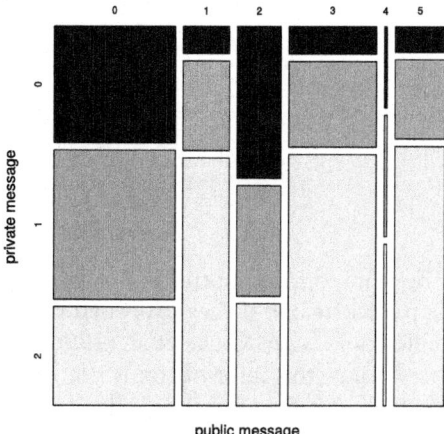

sages are more likely accompanied by more expensive public messages. A value of 4 (coalition and mobilization) is rarely used. The numbers indicate that coalition and mobilization are strategic substitutes, while massmedia are used to send less costly messages and frequently accompany one of the two. There is, however a fair share of groups using all available activities. Interest groups seem to balance costs and effects in their strategy choice, a point I will analyze further with the help of statistical models.

5.3   INDEPENDENT VARIABLES

My main goal is to test theoretical predictions which link individual characteristics and institutional constraints. While some variables are only measured on individuals and therefore independent from characteristics of the legislative events, others are event specific. Some event specific-variables vary by interest group.

The independent variables in my analysis are therefore measured at different levels. Individual level variables include the level of access, the lobbying costs, and the interest group type. Individual level variables which vary by event are the distance measure and the message costs. Variables measured at the event level are the uncertainty and the majority status of a bill. In the following, I discuss the variables (Table 5.3) and their operationalization in more detail.

Table 5.3
Measurement level of
independent variables

Group	Event	Group × Event
access	uncertainty	messaging costs
costs	bill type	distance
type		

5.3.1   Preferences and Distance Measures

The data contain information on whether the actors were in favor or against a particular bill, if they expressed concern with the bill. This binary variable codes what Knoke et al. (1996) call a preference. From a theoretical viewpoint, this information is not the appropriate preference measure. A distance measure based on a (latent) policy position is required to test the theory.

The theory builds on ideal points and the distance of these ideal points to ideal points of specific (potentially collective) decision-makers. Testing the model must rely on a distance measure which is to be derived from an estimate of the ideal points. The correlation between the ideal points and being in favor or against a bill is not high enough to use the latter as a proxy for the former. The theory also predicts that interest groups cannot prevent the change of a policy, simply being against the change of the policy is therefore a weak predictor of interest group activity.

To derive a measure of the latent ideal position, I exploit the fact that the preference data are structurally similar to roll call data. They can be used to calculate positions on a latent dimension. Knoke et al. (1996) and Pappi et al. (1995) used a factor analysis to identify latent positions of the actors. While this was a standard procedure at the time, today there are more advanced approaches at hand whose assumptions more closely match the assumptions of my theoretical model.

In addition, Pappi et al. (1995) used the preferences weighted by interests as a basis for the estimation of the positions. However, I will show that the interest variable captures a cost aspect which should not be lumped together with the position. One empirical approach to place interest groups and political decision-makers on a common policy space was proposed by McKay (2008). She combined roll call votes and survey data to estimate the positions using Nominate (Poole and Rosenthal 1985). My data are structurally equivalent to hers. I follow the general approach but I use the method proposed by Clinton et al. (2004) who use a Bayesian framework for the estimation of the latent policy positions. Their approach has two major advantages over Nominate.

Firstly, the Bayesian approach allows us to capture the degree of uncertainty which is particularly interesting in my setting, as there is a number of missing values caused by nonparticipation and self-selection into the events. Secondly, the model is set up based on the assumption of a quadratic utility loss function. In particular, it is assumed that the preferences fulfill

$$U_i(\zeta_j) = - \parallel x_i - \zeta_j \parallel^2 + \eta_{ij}$$

and

$$U_i(\psi_j) = - \parallel x_i - \psi_j \parallel^2 + v_{ij}$$

where x_i is the ideal point of actor i, and ζ_j and ψ_j are 'yes' and 'no' votes, respectively (Clinton et al. 2004, 356). An actor is assumed to vote in favor of a bill whenever his utility from adopting the bill is higher than the utility he derives from the Status Quo. The statistical model explicitly accounts for mistakes actors may make in the assessment of the situation.

Based on these assumptions, Clinton et al. (2004) suggest a procedure to estimate a latent policy position. This position is the latent factor which predicts the actual votes best. The dimensionality is determined by the researcher. As my model is explicitly modeled on a one-dimensional scale I used the approach to estimate latent policy positions on one policy dimension.

The specification of preferences used by Clinton et al. (2004) fits the assumptions of my theoretical model. This is in contrast to Poole and Rosenthal (1985) whose Nominate procedure is build on the assumption of Gaussian preferences. The functional form is quite important here as I am interested in distances. The assumed preference functions have very different implications for the utility derived when moving away from the ideal point. For quadratic utility loss functions the relative utility loss is strictly increasing. This implies that the punishment for accepting policies further away from one's ideal increases. For Gaussian preferences, the relative utility loss for moves away from the ideal point is increasing in the range between the inflection points of the curve. It declines for moves away from the idealpoint surpassing the inflection points. The assumptions which nominate is based on do therefore not match the assumptions of my theoretical model as Gaussian preferences imply a different type of utility model. The implications for the utility maximization are quite dramatic.

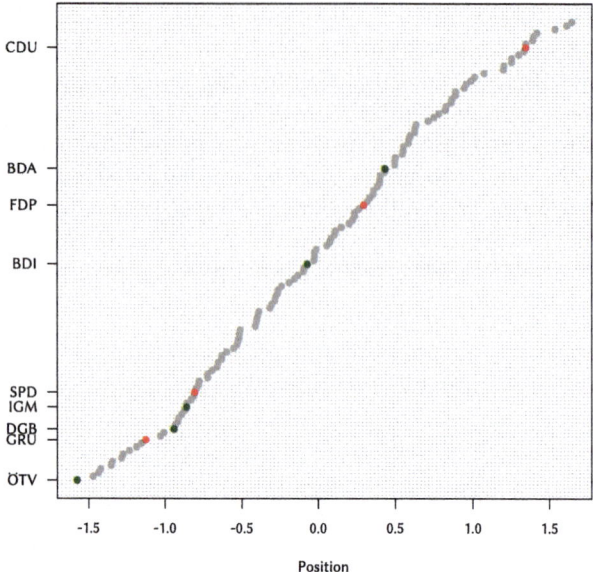

Fig. 5.5 Positions of political actors and interest groups in Germany (one-dimensional estimate)

Figure 5.5 shows the results of the estimation for all 126 actors in Germany. I have highlighted position estimates for the four parliamentary parties: The Greens (GRÜ), the Social Democratic Party (SPD), the Liberal Party (FDP), and the Christian Democratic Party (CDU). I also highlight some important interest associations: The Trade Union for Public Services and Transport (Ö-TV), the Trade Union for Steel Workers (IGM), the German Trade Union Association (DGB), the Association of the German Industry (BDI), and the German Employers' Association (BDA).

The location of the actors is in line with the expectations based on the general conflict structure in the policy field of labor and social policy in Germany. The Social Democrats, trade unions, and Greens can be found on one side of the policy dimension, the employers' associations and the governing coalition are located on the other side of the policy dimension.[7]

_____
[7] I have also estimated positions in a two-dimensional policy space. The face validity of these estimates is also good. However, as my theoretical model explicitly builds on one dimension I have chosen to run the analysis based on the one-dimensional estimate.

Table 5.4   Position and rank of political actors in the policy domain (Bundesrat and Bundestag)

Actor	ID	Position	Rank
Bremen	1104	−1.47	2
Hamburg	1105	−1.34	6
GRÜ-BT	915	−1.12	12
GRÜ-AS	916	−1.03	13
North Rhine-Westphalia	1108	−0.93	16
SPD-BT	908	−0.80	25
Hesse	1106	−0.69	31
SPD-AS	909	−0.51	40
Baden-Wuerttemberg	1101	−0.19	55
Saarland*	1110	−0.13	57
Berlin	1103	0.27	75
FDP-BT*	912	0.30	76
Schleswig-Holstein	1111	0.36	80
Rhineland-Palatinate	1109	0.40	83
FDP-AS	913	0.59	94
Lower Saxony	1107	0.82	102
CDU-AS	902	0.89	106
CDU-BT	901	1.35	119
Bavaria	1102	1.39	121

Asterisk indicates the median for each of the two chambers. The suffix BT identifies the party groups in the Bundestag, while the suffix AS identifies the party groups in the committee

Table 5.4 shows the estimated policy positions of the main political actors in the Bundestag and Bundesrat. For the former, these are the party factions and the committee factions. For the latter, these are the *Länder* governments. The median actor in the Bundesrat is the government of Saarland.

The median legislator can be used to represent the preference of the collective actor (Black 1958). Whenever I need to represent the position of the Bundesrat I use the position of the Saarland (−0.13) which is the median actor in the period. It is interestingly virtually identical to the vote weighted average of all positions (−0.136) in the Bundesrat. For the Bundestag, the median actor is the FDP. I use the position of the FDP to represent the position of the Bundestag whenever it is necessary.

The position of the government is a simple average of the positions of the two government parties (FDP and CDU) which is based on the assumption that both are equally strong as they can veto proposals of their coalition partner (Tsebelis 2002). Lastly, I calculate the distance between the two chambers as the distance between the government position and the position of the Bundesrat.

Distance to Center of Bargaining Range
The main explanatory variable for public messages is the distance to the expected policy outcome. The expected policy outcome is the midpoint of the bargaining range. The bargaining range depends on the majority status of the bill and the type of legislative procedure.

Opposition bills are seldom accepted by the government majority or a broad coalition of parties in Germany. This implies that from an interest group's viewpoint the process uncertainty is quite different for the two cases. In case of a majority bill the uncertainty may be considerable, while in the case of an opposition bill it is close to zero as the Status Quo will usually prevail. Higher process uncertainty implies higher interest group activity.

The case of Germany is particularly interesting for testing my theory because of the way the legislative process is structured. In some cases the bill needs approval by the Bundesrat ('Zustimmungspflicht'). In this case the legislative game is played between the two chambers, where both have a veto. If the second chamber does not have to consent, the game is played between the two coalition partners in the government coalition (Ismayr 2008). This implies that the location and size of the bargaining range differ between the two types of legislative procedure. As the decisive actors change from one process to the other the distances to the decisive actors or the center of the bargaining range change as well.

The distance measure is calculated as

$$centdist_{ik} = (-1)^j \cdot (C_k - X_i)$$

where X_i is the ideal position of actor i. The parameter j is one if the bill is a majority bill and zero if it is an opposition bill. The variable C_k is bill specific and determined by the type of legislative process. It is the mean position of the government if the law is not subject to approval by the Bundesrat. It is the midpoint of the distance between the Bundesrat and the Government position if the law is subject to approval by the Bundesrat.

These concepts are depicted in Fig. 5.6. The arrows indicate the value of the distance measure. Arrows pointing rightwards indicate positive distances, arrow pointing to the left indicate negative distances. The graphs depict the position of the interest group 'Bundesverband der Arbeiter-

Fig. 5.6 Distance to center of bargaining range as a function of veto type and majority status. Example for *Bundesverband der Arbeiterwohlfahrt (AWO)*

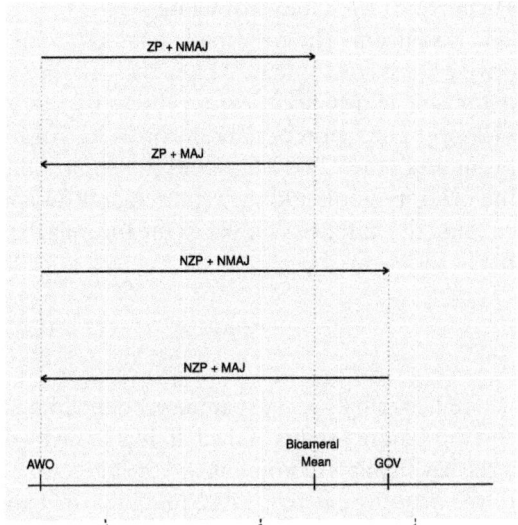

wohlfahrt (AWO)'. It was founded in 1919 by members of the Social Democratic Party (SPD) and was still part of the Social Democratic group of organizations surrounding the SPD and the major trade unions. Not very surprising, AWO is one of the most leftist groups in the sample. It has the fourth lowest idealpoint of all actors.

The distance measure for the AWO is negative in case that the bill at hand is a majority bill. A majority bill implies a rightward move on the policy dimension relative to the Status Quo. The expected policy outcome is therefore located to the right of AWO. In case of an opposition bill, the move relative to the Status Quo is leftward. The AWO therefore has a more progressive position. In order to treat the movements in the same way, the distance is measured on the inverted scale and hence positive. The cases are thus rendered comparable.

Taking the absolute value of the distance is not an option. Note that the equilibrium conditions of the theoretical model (e.g. condition 4.2) depend on the *directed* distance, not the absolute value of the distance. The operationalization of the distance variable has to reflect this. The reason is the asymmetry of the equilibrium partitions. Separating equilibria only exist in one part of the political space, in the other part only pooling equilibria exist.

Distance to the Compromising Actor

The distance to the expected policy outcome should not matter for the private message strategy. Here, the distance to the compromising actor matters. The distance measure shows more variation than the `centdist` measure. The theoretical model predicts that it is crucial to lobby the compromising actor. The identity of the compromising actor is determined by the majority status and the requirements for approval. The formula for calculating the distance s similar to the formula for the center to the bargaining range and is given as:

$$comdist_{ik} = (-1)^{j} \cdot (C_k - X_i)$$

where C_k is a bill specific parameter depending on the legislative procedure and the majority status of the bill. It is given by the position of the Bundesrat if the law requires its consent and the bill was introduced by the government or the majority parties. If it requires consent and is an opposition bill, C_k is the position of the Government. If the law does not require consent of the Bundesrat, C_k is the position of the FDP for majority bills and the position of the CDU for opposition bills. The parameter j is one if the bill is a majority bill and zero if it is an opposition bill. Figure 5.7 shows the coding

Fig. 5.7 Distance to compromising actor as a function of veto type and majority status. Example for *Bundesverband der Arbeiterwohlfahrt (AWO)*

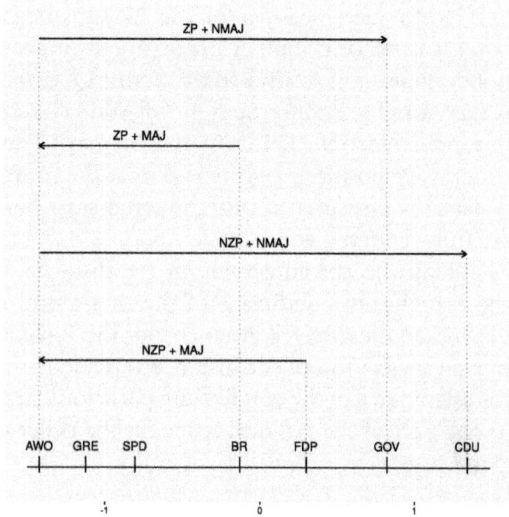

of this variable for the position of AWO graphically. The interpretation of the figure is analogous to Fig. 5.6. Rightward pointing arrows indicate positive distances, leftward pointing arrows indicate negative distances.

The compromising actor is crucial for determining the room for maneuver in the political process. The compromising actor will not accept policies which are further away from his ideal point than the Status Quo. How tight this restriction is in each individual case is impossible to analyze without knowledge about the exact location of the Status Quo. Unfortunately, it is impossible to empirically assess the exact location of the Status Quo based on the data . However, to derive hypotheses about the effect of the comdist variable on strategy choice the exact location of the Status Quo is not necessary. The distance measures and the direction of change of the Status Quo are sufficient.

5.3.2   *Lobbying Costs*

Lobbying costs are a major aspect of my theoretical model. They fall into two categories: (1) Interest groups face costs for gathering intelligence. The stock of knowledge is usually not readily accessible. Information has to be collected, compiled, and made intelligible in order to successfully communicate one's point. I term these costs *intelligence costs* and (2) Once information is gathered, it has to be transmitted. The interest group has to hire a lobbyist or send in-house lobbyists to meet decision-makers. One has to pay for a media placement or send out mailings to members. All these activities use resources like time and money and are therefore costly. These costs are named *message costs.*

It is impossible to directly assess the magnitude of the different costs in the dataset. I do, however, have measures at my disposal which allow me to make an approximation to these costs.

Intelligence Costs
The *Politikfeld Arbeit* project collected data on several types of networks in the domain of labor and social policy in Germany. The network on information flows is well-suited for the assessment of intelligence costs. While information flows potentially are a two-way street I focus only on the receiving part of the network. Participants were asked to name the domain actors they are receiving information from. Interest groups which receive information from many other groups are ceteris paribus expected to have lower information gathering costs. Interest groups with many incoming

connections are more likely to be well informed at a relatively low cost as others provide information to them.

Following Wasserman and Faust (1994, 179) I calculate a centrality measure as:

$$C_D = \frac{d(n_i)}{g - 1}$$

where $d(n_i)$ is the number of incoming connections for actor i and g is the total number of groups. The measure is standardized and gives the percentage of possible communication ties which are active for a specific actor. Note, however, that the measure does not vary across bills and thus is not capable to explain variation in behavior by itself. Nevertheless, controlling for this factor is important, as it may mask other effects driven by relative cost differences.

Message Costs

To assess the relative message costs requires a variable which is group specific but varies by event. A measure that is readily available is the interest, which is effectively a salience. I interpret the salience of a topic as indicating the relative costs of lobbying. Salience can be seen as the relative importance which an actor gives to an issue. The more important an issue for the member, the more 'salient' it is for the member's decision to act (cf. Hinich and Munger 1997, 52f.).

Hinich and Munger make the argument with respect to members of a committee, but it is applicable more generally. An important feature of salience is that it affects the shape of utility functions. If a dimension is more salient than another, then moves away from the idealpoint result in a bigger loss of utility. Salience is a relative concept, it makes only sense in a multidimensional space.

I assume that the policy space is one-dimensional. But the event space is multidimensional. The salience of the actor attaches a relative utility weight to each of the events. This allows me to use the salience as a measure of costs. A Euro spent on lobbying for an issue that is not salient signifies higher relative costs than a Euro spent on lobbying on a salient topic ceteris paribus. The relative utility loss (gain) compared to the Euro that is spent is the key component here. In my dataset the salience is measured by a variable ranging from one to five, where five indicates the highest salience.

In order to measure costs I take the inverse of the salience. The result is an indicator which ranges from 0.2 to 1. One indicates the highest relative costs in this case. The advantage of this formulation is that the interpretation is more straightforward, as a larger number implies larger *relative* costs.

$$C_m = \frac{1}{interest}$$

5.3.3   Fundamental Uncertainty

Fundamental uncertainty, as defined in my theoretical model, is a bill specific feature. A quantification will need to resort to observables, which makes it hard to measure the concept properly. I use two measures of fundamental uncertainty which provide variation across bills:

nPar This variable captures the number of paragraphs or articles in a bill. I have applied the following rules: If a bill consists only of paragraphs I have counted those. If a bill consists only of articles I have counted these. In bills that contained paragraphs nested in bills I have counted the articles which were not subdivided by paragraphs and added the count of paragraphs for the articles which were subdivided, not counting the article itself. The obligatory *Berlin-Klausel* which stated that the law would also be valid in West Berlin was not counted. Likewise, the purely procedural *Schlußbestimmungen*, which deal with topics like the enactment of a law were not counted.

nLawChange As so called 'Artikelgesetze', i.e. laws which contain articles usually modify several other laws, I have also coded a variable which counts the laws which are changed by the respective *Artikelgesetz*. As one article usually refers to one and only one other law, the count is in most cases identical with the count of articles. However, if there are articles which define a new law and thus are subdivided by paragraphs those counts differ.

Huber and Shipan (2002) use similar concepts to measure the complexity of bills. High complexity implies that it is hard to understand the problem at hand and to identify intended (and more importantly unintended) consequences of the bill on the table. Higher complexity therefore implies higher fundamental uncertainty.

Both measures are highly correlated and show similar patterns. A descriptive overview can be found in Table 5.5 together with descriptive statistics of the other independent variables.

Table 5.5 Summary statistics for independent variables

Variable	Minimum	1st quartile	Median	Mean	3rd quartile	Maximum	Missing
C_m	0.2	0.2	0.25	0.249	0.25	1	4
C_D	0.008	0.056	0.152	0.1785	0.264	0.72	16
ccost	0.2778	0.9191	1.6667	5.1945	5	62.5	20
nPar	1	3	5	10.85	18	43	0
nLawChange	1	2	3	6.814	9	26	0
nPage	1	3	7	9.861	9	44	0
BTaccess	1	2	3	3.196	4	8	80
BRaccess	0	1	2	2.297	3	8	0

5.3.4   Control Variables

Access

The studies of Pappi et al. (1995) and Knoke et al. (1996) elicited a variety of network relations in the policy field of labor and social policy. I focus on the general pattern of information flows. The receiving part of the network allows me to operationalize intelligence costs. But there is a second network, which is reported information sending. This sending network was created by surveying organizations about who they send information to. I use these data to derive a measure of access.

The general network structure maps all connections between actors without distinguishing between different institutional roles or powers of the actors. However, all actors are not created equal (Bräuninger and König 2004). I am mainly interested in access to the Bundesrat and the Bundestag who are the relevant political actors.

The measures are simple additive indices where the number of network contacts with either state governments (Bundesrat) or party factions/committee factions (Bundestag) is summed up. This gives two counts which I refer to as BTaccess and BRaccess (Fig. 5.8). The main interest is on the relative level of access to the chambers. This is important as theoretically the level of access to the compromising actor should matter much more.

The relative strength of the level of access is captured by calculating the percentage of access points in the Bundesrat relative to all access points in the two chambers for each interest group.

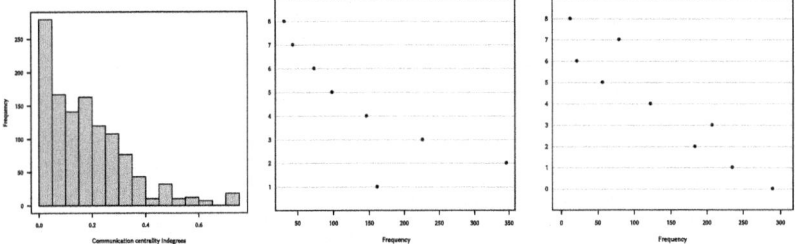

Fig. 5.8 Distribution of centrality indegrees, access to Bundestag and access to Bundesrat (left to right)

$$access = \frac{BRaccess}{BTaccess + BRaccess}$$

Majority Status

The political system of Germany is a parliamentary system. Looking at the success rates of opposition and government bills, it is evident that government bills are much more likely to to be accepted (Ismayr 2008). This creates different types of political uncertainty.

While an opposition bill is very likely to be voted down or die in committee, a majority bill is highly likely to become law. The majority status of a bill is, therefore, an important control variable. A dummy variable specification is used to capture this effect.

Type of Interest Group

One important aspect of interest groups is their type. The type allows a grouping of interest groups, for example, into business or professional societies. Some authors attribute interest group tactic choice and mobilization patterns to their type (Dür and Mateo 2013; Gais and Walker 2001). The type is determined by their general function in the system. Note that the group type in a conventional sense is fundamentally different from a type in a game theoretic sense. The two concepts should not be confused.

The 'Politikfeld Arbeit' study defined a set of types which were used to classify interest groups. The German classification was redefined by Knoke et al. (1996) to make it compatible with the US classification. I use the definition for the recoded variables as some types contain very few interest groups. The following types are used in the analysis: (1) Labor Unions and Work Councils, (2) Employers' Associations, (3) Professional Societies, (4) Public Interest Groups, (5) Ministries, (6) Political Parties, and (7) Länder (States).

5.4   HYPOTHESES REVISITED

We have now defined and operationalized all the relevant variables which we need to translate the hypothesis into empirical hypotheses.

Hypothesis 1 Interest groups with ideal points to the left of the expected policy outcome are, ceteris paribus, more likely to send public messages. The larger the distance, the higher is the likelihood of a public message.

- The effect of the variable `centdist` is negative.

Hypothesis 2 Interest groups with ideal points to the left of the ideal point of the compromising actor are, ceteris paribus, more likely to send private messages compared to interest groups with ideal points larger than the compromising actor. The larger the distance, the higher the likelihood of a private message.

- The effect of the variable `comdist` is negative.

Hypothesis 3 The more an interest group is in favor of a policy change (i.e., the more the group's ideal point is to the right), the more it, ceteris paribus, prefers private over public messages.

- The effect of the variable `comdist` is positive.
- The effect of the variable `centdist` is negative

Hypothesis 4 The higher the lobbying costs, the lower, ceteris paribus, the probability that an interest groups sends either public or private messages.

  a. The effect of the variable C_m is negative for both public and private messages.
  b. The effect of the variable C_D is positive for both public and private messages.

Hypothesis 5 Interest groups with relatively many access points in the Bundesrat are, ceteris paribus, more likely to send private messages.

- The effect of the variable access is positive.

Hypothesis 6 Higher fundamental uncertainty increases the likelihood that an interest group sends a message (either private or public).

   a. The effect of the variable nPar is positive for both public and private messages.
   b. The effect of the variable nLawChange is positive for both public and private messages.

REFERENCES

Black, D. (1958). *The Theory of Committees and Elections.* London: Cambridge University Press.

Bräuninger, T., & König, T. (2004). Senden und Empfangen: Informationstransfer in Politiknetzwerken als Vermittlung von Verhandlungsvorschlägen. In C. H. Henning & C. Melbeck, et al. (Eds.), *Interdisziplinäre Sozialforschung. Theorie und empirische Anwendungen* (pp. 205–224). Frankfurt: Campus Verlag.

Clinton, J., Jackman, S., & Rivers, D. (2004). The Statistical Analysis of Roll Call Data. *American Political Science Review, 98*(2), 355–370.

Coen, D. (2007). Empirical and Theoretical Studies in EU Lobbying. *Journal of European Public Policy, 14*(3), 333–345.

Dür, A., & Mateo, G. (2013). Gaining Access or Going Public? Interest Group Strategies in Five European Countries. *European Journal of Political Research, 52*(5), 660–686.

Esterling, K. M. (2004). *The Political Economy of Expertise: Information and Efficiency in American National Politics.* Ann Arbor: The University of Michigan Press.

Gais, T. L., & Walker, J. L. (2001). Pathways to Influence in American Politics. In J. L. Walker (Ed.), *Mobilizing Interest Groups in America,* Chapter 6 (pp. 103–121). Ann Arbor: University of Michigan Press.

Hinich, M. J., & Munger, M. C. (1997). *Analytical Politics.* Cambridge: Cambridge University Press.

Huber, J. D., & Shipan, C. R. (2002). *Deliberate Discretion? The Institutional Foundations of Bureaucratic Autonomy.* Cambridge: Cambridge University Press.

Ismayr, W. (2008). In W. Ismayr (Ed.), *Gesetzgebung in Westeuropa: EU-Staaten und Europäische Union* (pp. 383–430). Wiesbaden: VS Verlag für Sozialwissenschaften.

Jackman, S. (2008). Measurement. In J. Box-Steffensmeier, H. E. Brady, & D. Collier (Eds.), *Oxford Handbook of Political Methodology* (pp. 119–151). Oxford: Oxford University Press.

112   S. KOEHLER

Johnson, V. E., & Albert, J. H. (1999). *Ordinal Data Modeling*. Heidelberg and New York: Springer.

Knoke, D. (1990). *Political Networks: The Structural Perspective*. New York: Cambridge University Press.

Knoke, D., & Burleigh, F. (1989). Collective Action in National Policy Domains: Constraints, Cleavages, and Policy Outcomes. *Research in Political Sociology, 4,* 187–208.

Knoke, D., & Kaufmann, N. J. (1992). Social Organization of the United States National Labor Policy Domain [computer file]. Minneapolis, MN: David Knoke, University of Minnesota, Department of Sociology [producer]; Ann Arbor, MI: Inter-university Consortium for Political and Social Research [distributor]. doi:http://dx.doi.org/10.3886/ICPSR09802.

Knoke, D., & Pappi, F. U. (1991). Organizational Action Sets in the U.S. and German Labor Policy Domains. *American Sociological Review, 56*(4), 509–523.

Knoke, D., Pappi, F. U., Broadbent, J., & Tsujinaka, Y. (1996). *Comparing Policy Networks*. Cambridge: Cambridge University Press.

König, T. (1992). *Entscheidungen im Politiknetzwerk*. Wiesbaden: Deutscher Universitäts Verlag.

König, T., & Bräuninger, T. (1998). The Formation of Policy Networks. *Journal of Theoretical Politics, 10*(4), 445–471.

König, T., & Pappi, F. U. (1989). *Politikfeld Arbeit - Codebuch der Deutschen Teilstudie*. Kiel, Germany: Institut für Soziologie, Christian-Albrechts-University.

Laumann, E. O., & Knoke, D. (1987). *The Organizational State: A Perspective on National Energy and Health Domains*. Madison: University of Wisconsin Press.

Lowi, W. J. (1972). Four Systems of Policy, Politics and Choice. *Public Administration Review, 32*(4), 298–310.

Mahoney, C. (2007). Lobbying Success in the United States and the European Union. *Journal of Public Policy, 27*(1), 35–56.

McKay, A. (2008). A Simple Way of Estimating Interest Group Ideology. *Public Choice, 136,* 69–86.

Pappi, F. U., & Henning, C. H. C. A. (1998). Policy Networks: More Than a Metaphor? *Journal of Theoretical Politics, 10*(4), 553–575.

Pappi, F. U., König, T., & Knoke, D. (1995). *Entscheidungsprozesse in der Arbeits- und Sozialpolitik*. Frankfurt/M.: Campus Verlag.

Poole, K. T., & Rosenthal, H. (1985). A Spatial Model for Legislative Roll Call Analysis. *American Journal of Political Science, 29*(2), 357–384.

Shepsle, K. A., & Bonchek, M. S. (1997). *Analyzing Politics*. New York: W. W. Norton.

Tsebelis, G. (2002). *Veto Players: How Political Institutions Work*. Princeton: Princeton University Press.

Wasserman, S., & Faust, K. (1994). *Social Network Analysis*. Cambridge: Cambridge University Press.

CHAPTER 6

Interest Group Communication Strategies

Abstract In this chapter, I test the formal model empirically. I analyze the decision to mobilize and the decision to send public or private messages. I can demonstrate that the decision to send public messages depends on the distance to the expected policy outcome, while the decision to send private messages depends on the distance to the constraining actor. This is based on an identification strategy which uses the fact that the German political system functions as if the relevant decision-makers are exogenously assigned to issues based on the constitution.

Keywords Regression analysis · Germany · Lobbying strategy · Logistic regression

Many studies deal with the question of interest group influence. This is an inherently problematic endeavor (Dür 2008; Dür and de Bièvre 2007). Influence implies a change in behavior (Simon 1953). The fundamental problem is that assessing influence would require us to evaluate the counterfactual. We would need to analyze what policy would have been enacted, had the interest group not lobbied the decision-makers.

One example is the influence induced by strategic transmission of information in a separating equilibrium of the signaling/cheap talk games. Observable behavior following the exercise of influence is an equilibrium in a game-theoretic sense. Influence itself, however, is a process, not an equilibrium. The empirical problem is to draw inferences on influence by observing equilibrium behavior of the influenced. In this work, I therefore refrain from analyzing influence and try to asses whether the behavior of the

S. Koehler, *Lobbying, Political Uncertainty and Policy Outcomes*,
https://doi.org/10.1007/978-3-319-97055-4_6

interest groups who seek to exert influence can be explained as equilibrium behavior of an underlying strategic game.

I have argued that interest group communication strategies are comprised of two interrelated choices. The first choice is whether the interest group becomes active or not, i.e. mobilizes. The second choice is what course of action to take, given the decision to mobilize. I start with an analysis of the mobilization decision, before I move on to analyze strategy choice.

6.1   Mobilization Patterns

The first step in the analysis is to understand mobilization patterns. In the data at hand, there is a total number of 1204 event participants. The event participants are interest groups who were active in the event. The total number of potential participants is 32 (legislative events) · 126 (groups) = 4032. Hence, the average attendance rate is 29.9%. However, variation is quite large. Attendance can get as low as 11 groups (8.7%) and peaks at 92 groups (73.0%). The distribution of event participants is shown in Fig. 6.1. Many of the less well-attended events are opposition bills.

One argument often found in the literature on interest group tactics is that group type matters. Indeed, it is often seen as the most decisive factor. Disaggregating the attendance by group type shows interesting variation. Figures 6.2 and 6.3 depict the mobilization patterns by type of interest group. The former shows absolute numbers, the latter the percentage of groups within a category. At first sight, group type seems to have an independent effect on the decision to mobilize, as there is considerable variation across types of interest groups.

However, according to the theoretical model, mobilization should be driven by the ideological distance between the interest group and political actors, as it depends on the possibility of successful communication. The variation in the ideological spectrum within the different types of interest groups is of particular interest in this context. Figure 6.4 shows the distribution of actors' ideal points by group type. Variation of ideological positions is considerable, especially for the four types comprising nonpublic interest groups. It is likely that the ideological differences are independent factors which help explain event participation within groups. Failure to account for ideological proximity will put too much weight on group type as an explanatory factor. Indeed, as I will show later, group type rarely matters for

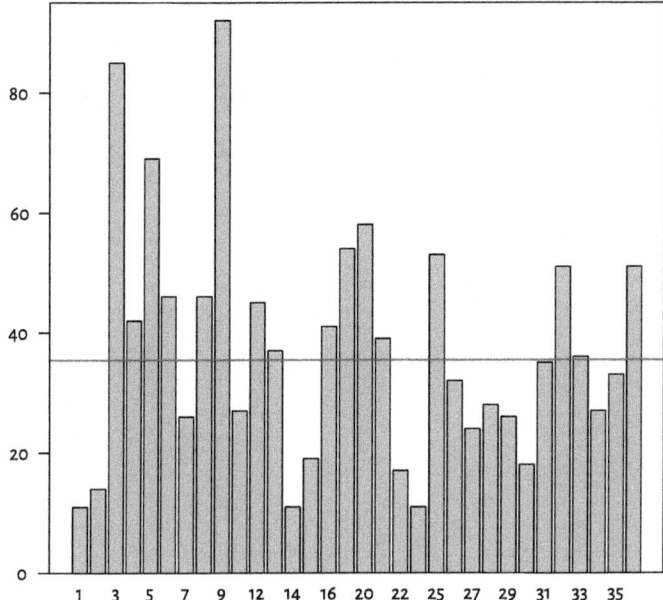

Fig. 6.1 Count of event participants. Red line indicates median number of participants per event

explaining interest group activity—once the ideological position of actors and the political process are taken into account.

6.1.1 Ideological Distance and Mobilization Patterns

To shed more light on the decision to mobilize, I added all the non-active groups to the dataset. This is easily done by creating the Cartesian product of the events and the groups. The extended dataset allows to draw inferences on the decision to mobilize. However, it is not possible to use regression analysis due to the selection problem. Recall that once an interest group indicated that they were not interested in an event they did not participate and hence where not asked any questions on the event.

This is equivalent to saying that lobbying costs were prohibitive for those groups. Running a regression to explain event participation on this newly created dataset would trivially find that the cost category created for those

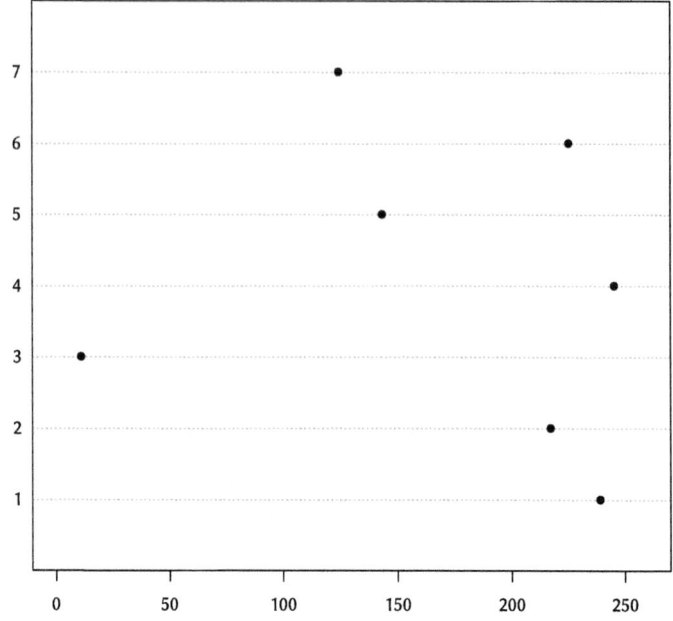

Fig. 6.2 Participation in events by group type (total count of active actor/event pairs)

cases perfectly predicts non-mobilization. This approach is therefore not very interesting.

At this point, it is important to note that the theoretical model postulates a nexus between costs and ideological differences. For a given cost, whether or not a group will mobilize will depend on the ideological distance to the expected policy outcome. Likewise, for a given ideological distance, the mobilization decision will depend on the level of costs. It is therefore possible to derive expectations about the patterns of the distribution of ideal points (distances) of active and inactive interest groups.

A second determinant for interest group strategy choice is the relative size of fundamental uncertainty and process uncertainty. For given fundamental uncertainty, process uncertainty is highest in case of a majority bill which is subject to approval by the Bundesrat as the bargaining range is larger. The lowest level of process uncertainty is the case of an opposition bill which is not subject to approval by the Bundesrat. These bills

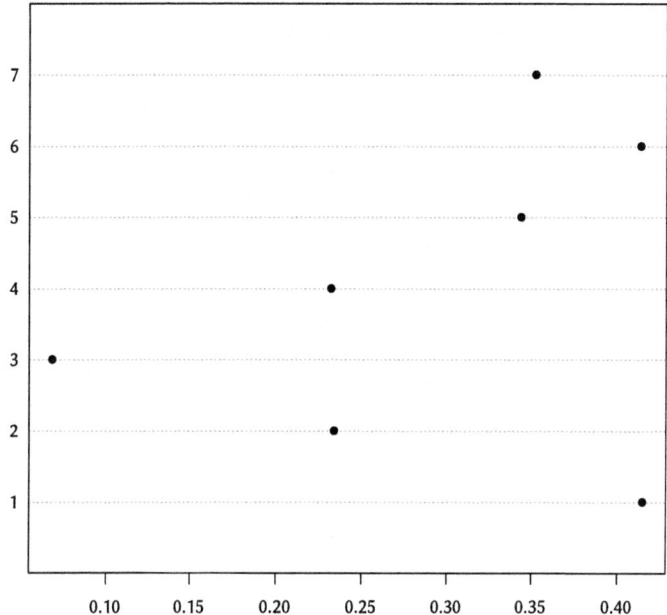

Fig. 6.3 Participation in events by group type (percentage of possible actor-event pairs)

are usually rejected by the parliamentary majority. The other two cases are intermediate cases. When the second chamber has a veto, the smaller bargaining range reduces the level of process uncertainty. When the bill is an opposition bill it has a higher likelihood to be accepted compared to the case of weak bicameralism. A majority bill has a higher likelihood of being accepted.

The distribution of the 126 domain actors' ideal points is shown in Fig. 6.5. Counting every group only one, it describes the spatial distribution of the ideal points. The distribution will serve as a useful reference for the analysis. Also, depicted is the empirical bargaining range for the two types of legislative procedures.[1] Bills which are subject to approval by the Bundesrat are indicated by ZP, NZP indicates that the bills which are not subject to approval.

[1] The bargaining range is sometimes also called the core.

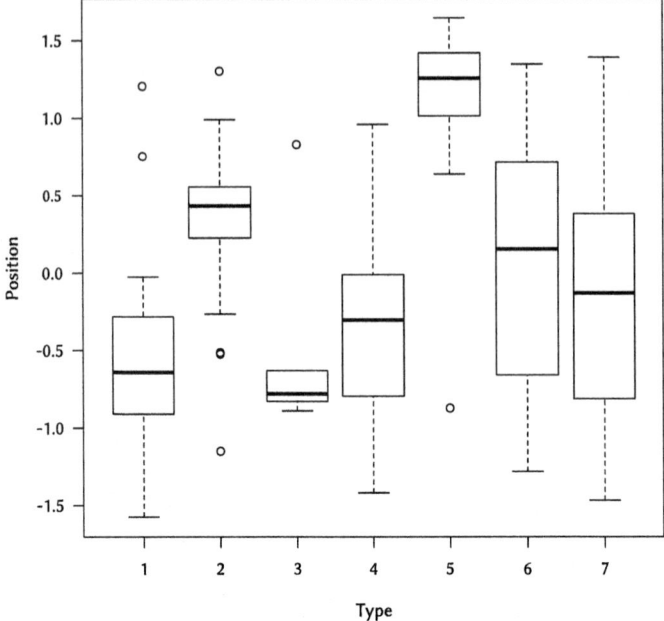

Fig. 6.4   Box and whiskers plots of distribution of actors ideal points by type

If the bill is subject to approval, the bargaining range is the distance between the position of the Bundesrat and the Government position. It is located closer to the center and is smaller compared to the cases where the bill is not subject to approval. In the latter case, the range is from the positions of the CDU to the position of the FDP.

A large part of the density is located between the two most extreme party factions, the Greens and the CDU. The position of the Bundesrat (represented by the position of the median country) is close to the center of the policy dimension. The highest density is located around the position of the FDP which was the median party in the Bundestag. Most of the interest groups are moderate relative to the spectrum of political parties. There are very few extreme interest groups.

Figure 6.6 shows the density for all groups (dotted line) in comparison to the density of ideal points for interest groups which mobilized, i.e. were active across all events (solid line). The latter density is more constrained

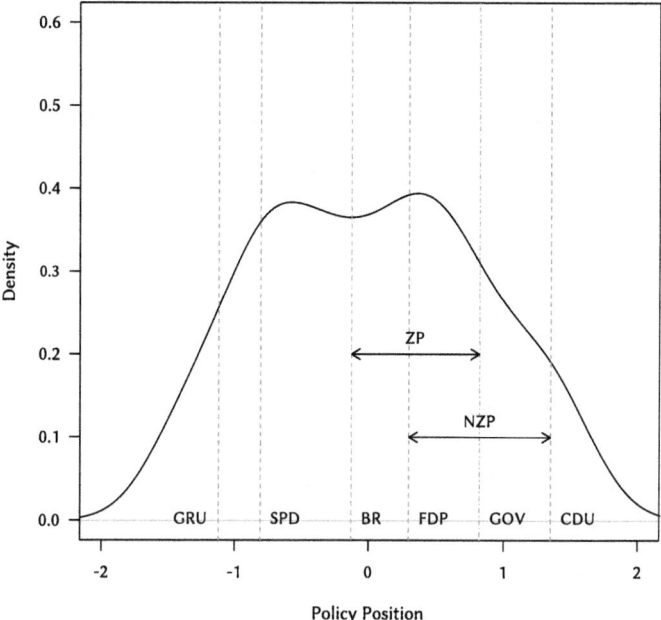

Fig. 6.5 Density of domain actors ideal points (all 126 actors)

from the right. In particular, the density of groups to the right of the center of the bargaining range is significantly smaller. There is a higher density of groups to the left of the center of the bargaining range. This is compatible with theoretical expectations of my model which argues that more extreme groups will not be able to communicate credibly and hence will not mobilize.

According to the theoretical model, the mobilization decision is determined by the ideological distance to the midpoint of the bargaining range, i.e. the expected policy outcome. Only interest groups to the left of this should be able to communicate credibly as long as communication is public. Figure 6.7 shows the density of the distance to the expected policy outcome for both active and inactive interest groups across all events. A negative distance implies that the ideal point is to the left of the expected bargaining outcome. The two densities are quite similar suggesting there may not be a difference. Figure 6.8 shows the comparison of the density

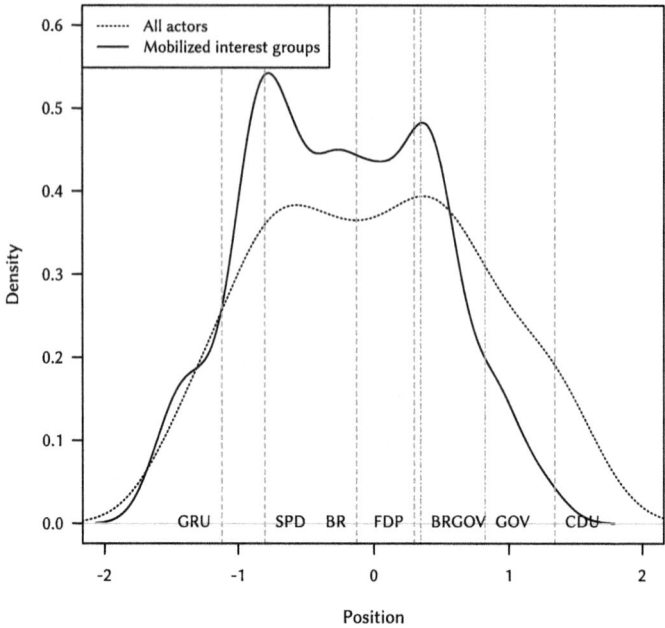

Fig. 6.6 Densities of mobilizing interest groups vs. general distribution

for active interest groups who do send public messages, compared with active interest groups which do not send public messages. The two densities cross exactly once, indicating they may not be the same. The density of the groups who send public messages seems to first order stochastically dominate the density for groups who do not send public messages. This is in line with the theoretical argument.

In order to put these impressions on a statistically sound basis, I use a two-sample Kolmogorov–Smirnov (K–S) test. It is easy to use the test to test for first-order stochastic dominance. The K–S test is a nonparametric test, well suited to compare two distributions which are *not* normally distributed and violate the assumptions of parametric tests like the t-test (Young 1977). As the underlying distributions, in this case, do not meet any of the assumptions required for the applicability of t-tests or other parametric tests, the K–S test is the tool of choice.

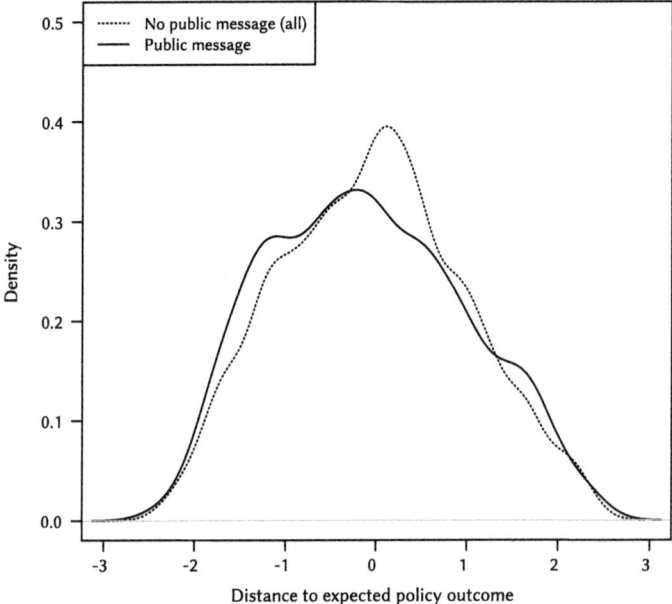

Fig. 6.7 Density of ideal points for active and inactive groups

The null hypothesis (H_0) of the K–S test states that the two samples are originating from the same underlying distribution. This can be more formally expressed as:

$$H_0 : f_1(x) = f_2(x)$$

I use the one-sided version of the test due to the clear expectation about the differences between distributions. The alternative hypothesis in this version of the test is that the distributions are different, more specifically

$$H_1 : f_1(x) > f_2(x)$$

which implies that the former is first-order stochastically dominated by the latter. I ran a K–S test for both comparisons. In the comparison between the active groups sending public messages $(f_1(x))$ and non-active $(f_2(x))$ groups we fail to reject the hypothesis that they are equal. In other words, there is no significant difference in the ideological distance to the center

122   S. KOEHLER

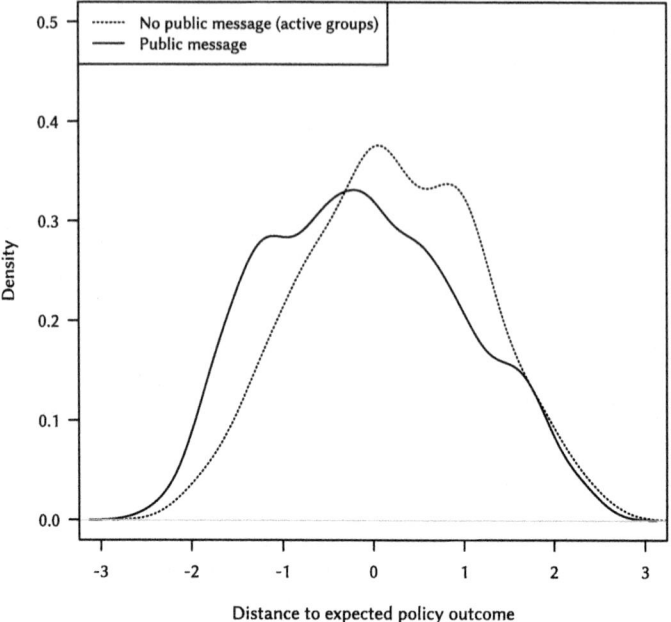

Fig. 6.8 Density of ideal points for active groups sending public messages and sending no message

of the bargaining range for active and inactive groups. Both groups are representative for the population as a whole.

In the case of the comparison for groups issuing public messages ($f_1(x)$) with the groups who are active but do not issue public messages ($f_2(x)$), one can reject the hypothesis that the distributions are equal ($D = 0.2065$, $p = 6.037 \cdot 10^{-7}$). The distribution of groups sending public messages is first-order stochastically dominated by the distribution of inactive groups. In other words, interest groups which send public messages are more likely to be to the left of the center of the bargaining range as the theoretical model would predict. They constitute a nonrepresentative sample of the population as a whole. This supports the idea that the mobilization is constrained by the bargaining environment and depends on the chance for successful communication.

6.2   Modeling Interest Group Strategies

The theoretical argument links institutional aspects (identity of veto players), the bill-specific context (majority/opposition bill) and individual choice behavior. Following the decision to mobilize, interest groups need to chose a lobbying strategy. According to the argument, the choice of strategy depends on the type of legislative procedure.

Several empirical hypotheses were derived. They are tested using the dataset on lobbying in Germany. The data structure can be understood as unbalanced panel data. It contains repeated measurements of interest group preferences, costs, and actions on predefined legislative events.

Major explanatory variables at the individual level are the message costs and most importantly, the positional distance. The first step is the analysis using models based on the standard assumptions of independent errors for each individual case and then move to the more appropriate cross-classified multilevel model. I demonstrate that the results are robust with respect to the choice of model and control variables.

The identification strategy rests on the exogenous variation in legislative procedure. Political actors do not have a choice on the use of legislative procedure. Articles 73 and 74 of the Grundgesetz determine the policy areas where the Bundesrat has a full veto or a suspensive veto, respectively. Due to this constitutional constraint, political actors are unable to choose the procedure strategically in order to gain political advantages.

In consequence, the legislative procedure is exogenous to the decision to legislate and the lobbying decision. Hence, the procedure provides identifying variation (Samii 2016) which I exploit in the statistical analysis.

6.2.1   Determinants of Strategy Choice

Model Choice and Approach

Interest groups' communication strategies can be comprised of public or private messages. I have derived several measures for the lobbying strategies. Theoretically, the resulting equilibria are partitioning equilibria, i.e. the cost-position space is partitioned into areas where communication occurs and areas where no communication occurs. In the statistical analysis, the cost-position space is assessed empirically.

For both public and private messages, the main dependent variable is binary. It codes whether interest groups chose a specific strategy or not. The variable is the closest possible to the theoretical model, where a message is binary signal. Strength of the message as a concept is meaningless in the model. The strength of the message can, however, be seen as a way to account for the relative costliness of the signal. The effects of the variables should therefore be more pronounced.

The test of my model rests on assumptions about actors utilities. First, it is true (by definition) that an actor chooses to send a public message whenever he is strictly better-off doing so. In other words, the utility from sending the message must be higher than the utility from not sending it. While theoretically convenient, this utility is not directly observable in empirical studies. There is, however, a way to make use of other variables to approximate the utility. The utility of an actor can be understood as a latent variable which determines the choice. The choice is directly observable. The underlying latent utility is approximated using other observables which may be alternative, context, or subject specific. The main focus turns to the analysis to what degree the observables help to explain observable choice behavior (e.g. Long 1997; Train 2009).

I analytically split the analysis of public and private messages at this point of the analysis. A logistic regression is the appropriate model to analyze the choice of the interest groups (e.g. Luce and Raiffa 1957; Train 2009). I use a series of models to test for the effects of explanatory and control variables on the probability of sending a private or a public message.

6.2.2   Basic Activities

The operationalization of the public and private strategies contains several individual activities. I now disentangle the effects of the specific activities in order to see if the patterns are similar and if not, which variables may drive the results. I use the same general determinants of interest group behavior as before.

The unit of analysis is the individual actors' decision to engage in a particular activity at a given legislative event. The two variables which form the private type of tactic are content and informal. The public tactic is constructed from the variables massmedia, mobilization and coalition. The results for the analysis of the individual activities are reported in Table 6.1. I also report the results for the variable formal. The results show empirically, that the variable is demand driven. The predictors for the activity are different from the other activities.

The results of the regressions for the individual activities show that for most activities, the predictors are comparable for activities classified similarly. This is further evidence that the activities can be reduced *classes*

Table 6.1 Regression results for individual activities

	Content	Informal	Formal	Massmedia	Mobilization	Coalition
(Intercept)	1.42**	1.33**	1.42***	1.11*	1.02†	−0.17
	(0.44)	(0.44)	(0.41)	(0.43)	(0.52)	(0.41)
distance (comdist)	0.24†	0.33*				
	(0.14)	(0.14)				
comdist · majority	0.12	−0.43*				
	(0.18)	(0.18)				
distance (centdist)			−0.06	0.67***	0.53**	0.40**
			(0.14)	(0.15)	(0.17)	(0.15)
centdist · maj			0.25	−1.28***	−1.02***	−0.83***
			(0.18)	(0.18)	(0.20)	(0.18)
majority	1.24***	1.19***	0.85***	0.62**	0.72**	0.45*
	(0.25)	(0.26)	(0.21)	(0.22)	(0.26)	(0.22)
C_m	−9.97***	−10.45***	−10.29***	−11.25***	−13.35***	−3.67**
	(1.30)	(1.32)	(1.33)	(1.44)	(1.82)	(1.22)
C_D	3.22***	5.51***	5.83***	3.16***	2.91***	−1.14*
	(0.56)	(0.64)	(0.65)	(0.52)	(0.48)	(0.47)
nPar	0.01	0.01	−0.01	0.02*	0.01	0.01
	(0.01)	(0.01)	(0.01)	(0.01)	(0.01)	(0.01)
access	0.30	−0.26	0.73*	0.61†	−0.11	−0.22
	(0.34)	(0.34)	(0.34)	(0.34)	(0.36)	(0.34)
N	988	1003	1004	1001	999	999
AIC	1120.70	1118.34	1105.67	1185.92	1104.28	1229.42
BIC	1277.36	1275.48	1262.85	1343.00	1261.30	1386.43
$\log L$	−528.35	−527.17	−520.84	−560.96	−520.14	−582.71

Standard errors in parentheses
† Significant at $p < .10$; $^{*}p < .05$; $^{**}p < .01$; $^{***}p < .001$

of functionally equivalent activities. This is encouraging as it will help to structure future research on interest group activities.

Despite the promising results, a note of caution is advisable. First, the interaction effects for the content and informal variables differ, albeit in inconsequential ways. The effect for both is in line with the expectations of Hypothesis 2. Second, the results for the predictors of the variable `formal` are in part deviating from the pattern for the public or private strategies. This may also be attributable to the fact that formal activities are to a large degree demand driven. Interest groups usually cannot nominate themselves for a hearing or membership in a workgroup. This seems to call for a different theoretical approach which is still to be developed. As to my knowledge, no theoretical approach is fully compatible with this finding.

Two things may be going on here at the same time. On the one hand, supply-driven interest group behavior seems to be compatible with the predictions of the model. On the other hand, demand-driven activities seem to be more in line with the resource exchange model of Hall and Deardorff (2006). For this line of research, a fundamental lack of formal models is evident. Developing more models along these lines to accompany approaches which analyze demand effects (e.g. Mahoney 2004) seems advisable. More effort is required to disentangle these effects both theoretically and empirically.

6.2.3   Public Messages

I use data on actual choices of the interest groups to draw inference on the determinants of lobbying tactics. A value of one signifies that the utility from sending a public message is higher than the utility from not sending a public message. The utilities are unknown and unobservable. But under the assumption that the utility can be modeled as a linear combination of independent variables, I can apply a logistic regression model to estimate the effect of these variables on the group's choice of activity.

The major explanatory variable used to predict behavior is the distance to the compromising (collective) actor or the expected policy outcome, depending on the type of strategy. Following the arguments of the theoretical models, it is the distance that matters and not the absolute position of the interest group.

A second set of important variables is the lobbying costs. They determine for which distances lobbying pays for the interest groups and when lobbying becomes too costly. The distance to the expected policy outcome and the two cost measures are the variables I use to predict the choice of public messages. The effect of the distance measure depends on the majority status of the bill.

It is very unlikely that an opposition bill is accepted. As a result, lobbying may be guided by different utility calculations. I, therefore, add an interaction with the dummy variable which codes majority status. I start with this parsimonious regression model and then add more control variables in subsequent regressions. The general structure of the empirical models is

$$Pr(public) = logit^{-1}(\beta_0 + \beta_1 centdist + \beta_2 majf + \beta_3 centdist \cdot majf$$

$$+\beta_4 C_m + \beta_5 C_D + \ldots)$$

where the dots signify the additional variables I control for. The main effect of interest is the effect of the distance to the expected policy outcome (or the midpoint of the bargaining range). The theoretical model predicts this effect to be positive (see Hypothesis 1).

Table 6.2 shows the results of the regression models. The estimated effect of the distance to the center of the bargaining range (centdist) is positive. This is in contrast to what was predicted by the theoretical model. However, the interaction term is significant and negative. Taking the interaction into account, the effect of the distance is indeed negative for majority bills but not for opposition bills.

In Figs. 6.9 and 6.10 the effect of the distance to the center of the bargaining range for different levels of costs is graphed. The y-axis displays the predicted probability of a message as a function of the distance to the expected policy outcome, holding the other variables at their means or a baseline level for factors, such as group type. The graphics on the left display the effect for majority bills, while the graphics on the right display the effect for opposition bills. The different lines sketch the effect for different values of lobbying costs. Generally, higher lines imply lower costs. In the top row the messaging costs are varied, in the bottom row intelligence costs are varied. The calculation of the predicted probabilities is based on the fully specified model (4) from Table 6.2.

The effect of the distance to the center of the bargaining range is different for majority and opposition bills, which is a striking fact in light of the formal model. While the overall effect is in line with theoretical expectations for majority bills, it is reversed for opposition bills. This suggests that the theoretical model does a much better job in predicting interest group behavior regarding bills introduced by the government. Given that this is the vast majority of bills issued in parliamentary regimes the model helps to understand a good deal of interest group activities.

In order to understand interest group behavior regarding opposition bills a different approach is necessary. The effect observed for the opposition bills rather suggests that interest groups use public strategies to signal not to decision-makers but to their own constituency. Understood this way, the findings suggest a complementary approach to understanding private and public messages based on the legislative procedure and the corresponding goals of interest groups. It seems that interest groups engage in group maintenance activities, when their behavior does *not* matter for outcomes. However, a fully integrated theory would require a fundamentally different

Table 6.2   Results of logit analysis for public messages

| | Dependent variable: Public message | | | | |
	Model 1	Model 2	Model 3	Model 4	Model 5
(Intercept)	1.45***	1.15**	0.99*	1.33**	1.38**
	(0.34)	(0.35)	(0.40)	(0.47)	(0.47)
distance (centdist)	0.77***	0.78***	0.81***	0.77***	0.77***
	(0.14)	(0.14)	(0.15)	(0.18)	(0.18)
majority	0.82***	0.75***	0.78***	0.85***	0.76**
	(0.19)	(0.19)	(0.20)	(0.23)	(0.24)
distance · majority	−1.50***	−1.53***	−1.59***	−1.54***	−1.53***
	(0.18)	(0.18)	(0.19)	(0.25)	(0.25)
C_m	−8.41***	−7.79***	−7.49***	−7.74***	−7.52***
	(1.16)	(1.17)	(1.23)	(1.30)	(1.31)
C_D	1.90***	1.96***	2.36***	1.82**	1.73**
	(0.49)	(0.51)	(0.54)	(0.65)	(0.65)
nPar		0.02**	0.02**	0.02**	
		(0.01)	(0.01)	(0.01)	
nLawChange					0.03*
					(0.01)
access			−0.22	0.30	0.29
			(0.34)	(0.42)	(0.42)
Business				−0.05	−0.08
				(0.29)	(0.29)
Professional Soc.				−2.64*	−2.76*
				(1.28)	(1.30)
Public IG				−0.71**	−0.77**
				(0.26)	(0.26)
Ministries				−1.13**	−1.16***
				(0.35)	(0.35)
Parties				1.18***	1.18***
				(0.35)	(0.35)
States				−2.04***	−2.06***
				(0.33)	(0.33)
N	1138	1068	992	992	992
AIC	1305.11	1217.95	1140.87	1010.50	1011.89
BIC	1426.00	1357.21	1297.66	1284.89	1286.27
log L	−628.56	−580.97	−538.43	−449.25	−449.94

Standard errors in parentheses
[†] Significant at $p < .10$; *$p < .05$; **$p < .01$; ***$p < .001$

modeling approach. Developing such a combined model for both activities would be an interesting extension for future work.

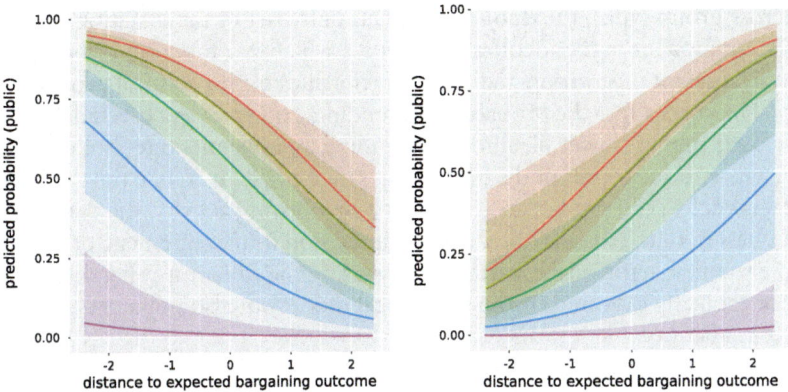

Fig. 6.9 Predicted probability of public message as a function of the distance to the expected policy outcome for various levels of message costs for majority bills (left) and opposition bills (right). Higher curves imply lower costs

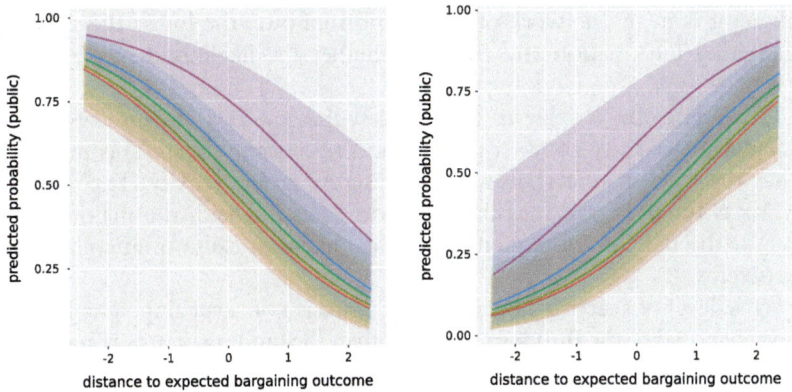

Fig. 6.10 Predicted probability of public message as a function of the distance to the expected policy outcome for various levels of intelligence costs for majority bills (left) and opposition bills (right). Higher curves imply lower costs

It is important to highlight that this effect is still observable after controlling for group type. As group type is often considered a major predictor of interest group activities, this finding is particular interesting. While some of the group dummies are significant, indicating differences in strategy choice across group types, the dummies explain only part of the variation. A major component is the ideological distance which is a function of both the individual interest group and the characteristics of the political process.

The costs affect the probability of sending a message as expected. Higher costs imply a lower likelihood of sending a public message, holding the distance to the center of the bargaining range constant. While the theoretical model would predict an abrupt drop in lobbying activity, the data show a phasing out. This may be due to imprecision in measurement or due to the fact that actors make mistakes in assessing the situation at hand. Both is not explicitly captured by the theoretical model but also not a problematic contradiction.

The two types of lobbying costs have the expected signs. Both effects are consistent and robust across model specifications and in line with the theoretical expectation as highlighted in Hypothesis 5. The effect of the messaging costs is negative and significant. The more an interest group cares about the issue, the more it is willing to send costly public messages. The effect of the intelligence costs is positive. That is, the more central an actor is in the network of information inflows, the lower the costs of gathering information and thus, the higher the likelihood of sending a message.

Including the variable nPar or nLawChange as a measure of fundamental uncertainty shows a positive and significant effect. This indicates that the level of uncertainty affects interest group activity. The variable *nPar* is the preferred measure as it correlates less with the majority status of the bill. In consequence, it is less likely to cause multicollinearity problems.

The level of access is not significant. This is not surprising. The relative level of access to the Bundesrat should affect the sending of private messages but not the sending of public messages, as access is not required to engage in public communication. Public communication will still be targeting the decision-makers, rather than the public or the interest groups' constituency.

6.2.4   *Private Messages*

The empirical models used for the analysis of private messages are very similar to the models used for public messages. The major difference is that the distance relevant for the choice of message is the distance to the compromising actor. This follows from the theoretical model, in particular, Eq. (4.15). All models include the distance measure represented by the variable `comdist`. Otherwise, they are structurally identical to the model for public strategies. This is a necessity, given that strategy choice is taking place in the same environment. A general representation of the model is:

$$Pr(private) = logit^{-1}(\beta_0 + \beta_1 \cdot comdist + \beta_2 \cdot comdist \cdot majority + \beta_3 \cdot majority$$

$$+\beta_4 \cdot C_m + \beta_5 \cdot C_D + \ldots)$$

where the dots represent additional control variables I subsequently add to the model (Table 6.3).

The variable `comdist` which measures the distance to the compromising actor is significant in all regressions. In line with the theoretical model, the effect is positive. Figures 6.11 and 6.12 show the predicted probabilities. The results for this regression are not in line with Hypothesis 2. They are, however, compatible with Hypothesis 3. This is further evidence that the choice of strategy is dependent on the costs and the distance in a given legislative process. Thus, one would expect a positive effect of the `comdist` variable if one takes into account the interdependence of the strategy choice between public and private messages.

It is important to note that the interaction effect with the majority dummy is not significant. Interest groups do not change the behavior dependent on the type of procedure. This seems logical, as private messages cannot be used to signal to the constituency. Therefore, the findings are in line with the theoretical model.

The other effects are as expected. Higher message costs come with a reduced probability of sending a private message. The result holds for both messaging and intelligence costs. The cost structure affects private and public message in a similar way. This was as was to be expected, as we only measure relative costs. They should therefore affect both choices in qualitatively similar ways.

Fundamental uncertainty affects private messages in a fashion similar to public messages. However, the effect is not significant. Particularly interesting is the `access` variable. Theory would predict that only the compromising actor is targeted if fundamental uncertainty is not too high. In most cases, this is the Bundesrat. Actors with relatively good access to the Bundesrat should therefore send more private messages. The positive and significant effect for the `access` variable underlines this effect.

Table 6.3 Results of logistic regression for private messages

| | Dependent variable: Private message | | | | |
	Model 1	Model 2	Model 3	Model 4	Model 5
(Intercept)	2.19***	2.14***	1.52**	0.72	0.76
	(0.41)	(0.42)	(0.47)	(0.54)	(0.54)
distance	0.28†	0.28†	0.25†	0.59**	0.59**
	(0.15)	(0.15)	(0.15)	(0.19)	(0.19)
majority	1.66***	1.63***	1.67***	2.13***	2.05***
	(0.26)	(0.27)	(0.28)	(0.32)	(0.33)
distance · majority	0.22	0.12	0.18	−0.40	−0.40
	(0.20)	(0.20)	(0.21)	(0.28)	(0.28)
C_m	−10.76***	−10.85***	−10.04***	−10.59***	−10.46***
	(1.26)	(1.30)	(1.36)	(1.42)	(1.43)
C_D	5.40***	5.37***	4.90***	5.07***	4.99***
	(0.78)	(0.80)	(0.82)	(0.88)	(0.88)
nPar		0.01	0.01	0.01	
		(0.01)	(0.01)	(0.01)	
nLawChange					0.03
					(0.02)
access			1.29**	1.70***	1.67***
			(0.41)	(0.50)	(0.50)
Business				0.67*	0.66*
				(0.34)	(0.34)
Professional Soc.				0.01	−0.06
				(1.37)	(1.37)
Public IG				0.28	0.23
				(0.29)	(0.29)
Ministries				1.26**	1.23**
				(0.46)	(0.46)
Parties				0.28	0.28
				(0.32)	(0.32)
States				−0.53	−0.55
				(0.38)	(0.38)
N	1113	1043	973	973	973
AIC	917.64	860.09	800.99	797.66	797.78
BIC	1038.00	998.69	957.17	1070.96	1071.09
log L	−434.82	−402.05	−368.50	−342.83	−342.89

Standard errors in parentheses
† Significant at $p < .10$; *$p < .05$; **$p < .01$; ***$p < .001$

Many interest groups who sent a private message also sent a public message. As there is a large degree of overlap in the positions for which the messages are possible this is not surprising. Moreover, from a theoretical

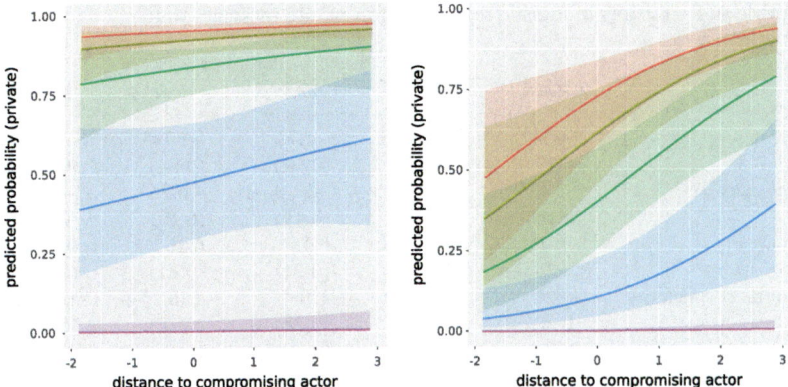

Fig. 6.11 Predicted probability of private message as a function of the distance to the compromising actor for various levels of message costs for majority bills (left) and opposition bills (right). Higher curves imply lower costs

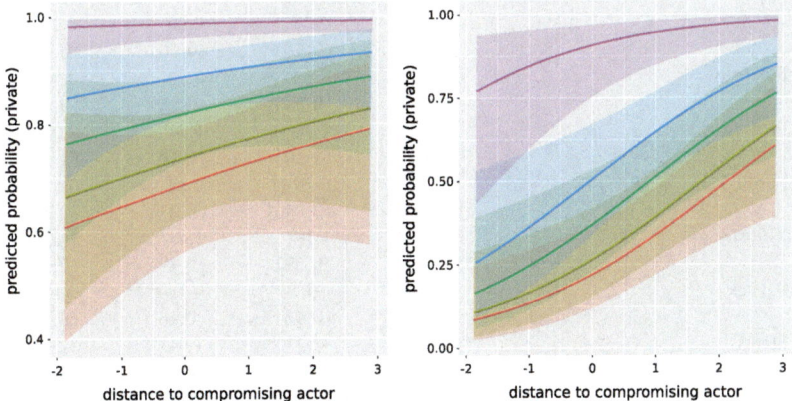

Fig. 6.12 Predicted probability of private message as a function of the distance to the compromising actor for various levels of intelligence costs for majority bills (left) and opposition bill (right). Higher curves imply lower costs

viewpoint, the public message is the more effective message as it informs both chambers. To get a more precise idea about interest group activities, I ran a statistical model where the dependent variable is one only for interest

Table 6.4 Results of logistic regression for private messages only

	Private messages only				
	Model 1	Model 2	Model 3	Model 4	Model 5
(Intercept)	−0.81*	−1.78***	−1.78***	−1.78***	−1.85***
	(0.39)	(0.54)	(0.54)	(0.54)	(0.53)
distance	−0.42**	−0.26	−0.26	−0.26	−0.26
	(0.16)	(0.19)	(0.19)	(0.19)	(0.19)
majority	−0.18	0.15	0.15	0.15	0.19
	(0.26)	(0.31)	(0.31)	(0.31)	(0.31)
distance · majority	1.32***	1.10***	1.10***	1.10***	1.09***
	(0.20)	(0.28)	(0.28)	(0.28)	(0.28)
C_m	0.74	0.63	0.63	0.63	0.56
	(1.05)	(1.18)	(1.18)	(1.18)	(1.19)
C_D	−1.21*	−0.42	−0.42	−0.42	−0.35
	(0.55)	(0.69)	(0.69)	(0.69)	(0.69)
nPar	−0.02*	−0.02*	−0.02*	−0.02*	
	(0.01)	(0.01)	(0.01)	(0.01)	
nLawChange					−0.02†
					(0.01)
access		0.28	0.28	0.28	0.29
		(0.46)	(0.46)	(0.46)	(0.46)
Business		0.54	0.54	0.54	0.56†
		(0.33)	(0.33)	(0.33)	(0.33)
Prof. Soc.		1.94†	1.94†	1.94†	2.03†
		(1.10)	(1.10)	(1.10)	(1.11)
Public IG		0.51	0.51	0.51	0.57†
		(0.31)	(0.31)	(0.31)	(0.32)
Ministries		1.18**	1.18**	1.18**	1.22**
		(0.39)	(0.39)	(0.39)	(0.39)
Parties		−1.54***	−1.54***	−1.54***	−1.54***
		(0.46)	(0.46)	(0.46)	(0.46)
States		1.94***	1.94***	1.94***	1.97***
		(0.37)	(0.37)	(0.37)	(0.37)
N	1050	976	976	976	976
AIC	1051.62	874.97	874.97	874.97	876.17
BIC	1190.40	1148.45	1148.45	1148.45	1149.64
$\log L$	−497.81	−381.49	−381.49	−381.49	−382.08

Standard errors in parentheses
†Significant at $p < .10$; *$p < .05$; **$p < .01$; ***$p < .001$

groups who *exclusively* sent a private message. The results for these models are reported in Table 6.4.

The results differ from the general results for private messages. With the exception of one model, the effect of the distance to the compromising actor is no longer significant. However, the interaction effect with the majority dummy is highly significant in this setup. The effect of the distance is therefore positive, for majority bills and negative for opposition bills. This is interesting as this indicates that groups who do only send private messages behave differently, depending on the type of legislative procedure. In particular, for majority bills, the groups who resort to sending only private messages are more likely in favor of the policy, while for opposition bills, they are more likely to be against a policy.

Interestingly, the messaging costs are not significant in this setup. The actors seem to send the messages irrespective of the costs involved. However, if one acknowledges the fact that public messages seem more effective *and* are more expensive, this could indicate that some interest groups who care less, still care enough to be willing to pay the price of a private message. An alternative explanation would be that messaging costs of private messages are empirically close to zero which would mean that they are indeed sending messages to the accommodating chamber. Unfortunately, the data do not allow me to empirically disentangle these effects.

The level of uncertainty has the opposite effect, it is more likely that interest groups send only private messages in case of low uncertainty. It is interesting to note that the access variable is insignificant. The level of access does not help to predict private messages in this case. This is a strong indication that public messages are not a fallback option for groups without access. The finding is in line with Binderkrantz (2008).

Appendix

Multilevel Models
Standard regression models assume that the errors of individual cases are independent from each other (e.g. Wooldridge 2002). In the dataset at hand, two measurements for a single interest group are potentially not fully independent from each other. The interest group may have a general predisposition to use one activity over the other—independent from the context. This assumption is the basis for studies who attribute the choice of activities to interest group characteristics (e.g. Gais and Walker 2001). The error terms for individual actors may therefore be correlated, thereby violating a major assumption of many statistical models, although this would be mitigated by the fact that the panel is unbalanced.

In addition, more than one interest group participates in the events. It is unlikely that the cases within an event are fully independent from each other. Certain contexts may cause actors to use specific actions so the correlation of interest groups tactics and therefore error terms for a given event is a possibility. For example, coalitions are a factor which influences interest groups' choice at a given point in time. The resulting correlation of errors may be problematic for the empirical modeling of strategies as well.

Using a standard regression model which assumes independent errors on each observation is therefore not necessarily appropriate. A data structure where a serious correlation of errors for specific groups and across events is possible could be problematic. The structure of (potential) error correlation calls for a non-nested hierarchical model with varying intercepts. This type of model accounts for the interdependence of the error terms (Gelman and Hill 2007).

The multilevel modeling approach has another advantage. For the application of the model the data need not be balanced (Hox 2010, 79). This is particularly important as the dataset is an unbalanced panel. Not all interest groups participate in all legislative events. Biases caused by nonattendance are therefore not a problem in the multilevel models.

The general formula for the estimated models is

$$y_i \sim logistic(X_i\beta + \eta_{j[i]} + \mu_{k[i]})$$

$$\eta_{j[i]} \sim N(0, \sigma_\eta^2)$$

$$\mu_{k[i]} \sim N(0, \sigma_\mu^2)$$

where $\eta_{j[i]}$ denotes the effects for interest group j and $\mu_{k[i]}$ represents the effect for event k. I estimate these models using the glmer function of the lme4 package in R (Bates et al. 2013). Note that because I am now estimating the effect of individual events, I can no longer include the majority dummy in the model. Non-varying variables at the individual level can also no longer be included.

The effects are qualitatively comparable to the results of the standard logistic regressions. The distance to the expected policy outcome is negative and significant. The costs and the uncertainty have the expected effects.

Explicitly modeling the interdependencies in the error structure seems not to dramatically change the results. This suggests that the simple logistic regression models capture the essence of the situation and are sufficient for

Table 6.5   Results of cross-classified hierarchical logistic regression

	Public messages			
	Model 1	*Model 2*	*Model 3*	*Model 4*
(Intercept)	7.33*	6.71*	5.84*	6.03*
	[5.97; 8.69]	[5.32; 8.10]	[4.20; 7.48]	[4.41; 7.65]
distance	−0.31*	−0.34*	−0.35*	−0.38*
	[−0.58; −0.03]	[−0.62; −0.07]	[−0.63; −0.06]	[−0.68; −0.08]
C_m	−18.18*	−17.50*	−17.01*	−17.29*
	[−22.29; −14.06]	[−21.73; −13.27]	[−21.47; −12.55]	[−21.72; −12.86]
access			2.09	2.08
			[−0.10; 4.29]	[−0.10; 4.26]
nLawChange		0.08*	0.08*	
		[0.02; 0.13]	[0.02; 0.14]	
nPar				0.04*
				[0.01; 0.07]
AIC	754.05	712.54	665.15	667.57
BIC	779.30	742.47	699.55	701.98
Log Likelihood	−372.03	−350.27	−325.57	−326.79
Num. obs.	1153	1083	1007	1007
Num. groups: OrgID	126	125	112	112
Num. groups: EventID	32	28	28	28
Var: OrgID (Intercept)	4.16	4.02	4.15	4.09
Var: EventID (Intercept)	0.55	0.29	0.33	0.34

95% confidence interval reported below coefficients
*0 outside the confidence interval

analyzing interest group behavior. This also suggests that strategy choice is neither driven by group characteristics (type) nor by context alone, but by an interaction of the two components (Table 6.5).

The results for the multilevel regressions on private messages are similar to the standard logistic regression. The model also predicts a positive effect of the distance to the compromising actor. The costs have the expected sign and also the uncertainty measure has a comparable effect. One interesting difference is the insignificance of the access term. This is most likely driven by the decomposition into individual events (Table 6.6).

Alternative Operationalization of the Dependent Variable
As a last robustness check, the strength of the signal is accounted for in the dependent variable. The variation I report here is the simple count of the activities. The model of choice for this type of dependent variable is a negative binomial regression model. Table 6.7 shows the results for a

Table 6.6   Results of cross-classified hierarchical logistic regression

	Private messages			
	Model 1	Model 2	Model 3	Model 4
(Intercept)	6.78*	6.24*	5.14*	5.55*
	[5.31; 8.25]	[4.62; 7.85]	[3.31; 6.98]	[3.70; 7.41]
distance	0.46*	0.40*	0.43*	0.43*
	[0.19; 0.74]	[0.12; 0.68]	[0.15; 0.71]	[0.15; 0.72]
C_m	−15.52*	−15.47*	−15.26*	−15.61*
	[−19.61; −11.42]	[−19.87; −11.07]	[−19.77; −10.74]	[−20.14; −11.09]
nPar				0.05
				[−0.01; 0.10]
nLawChange		0.13*	0.13*	
		[0.04; 0.22]	[0.04; 0.23]	
access			2.20	2.20
			[−0.41; 4.80]	[−0.42; 4.82]
AIC	785.92	724.78	693.33	697.50
BIC	811.07	754.57	727.61	731.77
Log Likelihood	−387.96	−356.39	−339.67	−341.75
Num. obs.	1129	1059	989	989
Num. groups: OrgID	126	125	112	112
Num. groups: EventID	32	28	28	28
Var: OrgID (Intercept)	6.25	6.97	5.91	6.00
Var: EventID (Intercept)	1.63	1.26	1.45	1.77

95% confidence interval reported below coefficients
*0 outside the confidence interval

simple count of public alternatives used. The results for the count of private messages can be found in Table 6.8.

The results are very similar to the results of the standard logistic regressions. Particularly interesting is that the actors for which the distance is larger are more likely to send a strong (and thereby costly) public message. This is in line with the predictions of the model, that the range of costs for which an equilibrium exists is larger, if the distance to the expected policy outcome is large. The results for the private messages are also in line with the results of the simple logistic regression. The simple binary operationalization captures all relevant aspects of the strategic interaction.

Table 6.7   Results of negative binomial regression for count of public activities

	Model 1	Model 2	Model 3	Model 4	Model 5
(Intercept)	1.00***	0.92***	0.82***	0.91***	0.93***
	(0.18)	(0.18)	(0.21)	(0.22)	(0.22)
distance	0.31***	0.31***	0.32***	0.26***	0.26***
	(0.07)	(0.07)	(0.07)	(0.08)	(0.08)
majority	0.39***	0.38***	0.37***	0.35**	0.34**
	(0.10)	(0.10)	(0.11)	(0.11)	(0.11)
distance · majority	−0.55***	−0.56***	−0.58***	−0.48***	−0.48***
	(0.08)	(0.08)	(0.08)	(0.10)	(0.10)
C_m	−6.00***	−5.79***	−5.60***	−5.42***	−5.39***
	(0.63)	(0.64)	(0.68)	(0.69)	(0.69)
C_D	0.66***	0.68***	0.72***	0.26	0.24
	(0.17)	(0.17)	(0.17)	(0.18)	(0.18)
nPar		0.00†	0.01*	0.00	
		(0.00)	(0.00)	(0.00)	
access			0.07	0.10	0.09
			(0.15)	(0.18)	(0.18)
Business				0.10	0.09
				(0.11)	(0.11)
Prof. Soc.				−0.99	−1.03
				(0.71)	(0.71)
Public IG				−0.15	−0.17†
				(0.10)	(0.10)
Ministries				−0.68***	−0.69***
				(0.17)	(0.17)
Parties				0.43***	0.43***
				(0.09)	(0.09)
States				−0.53***	−0.54***
				(0.14)	(0.14)
nLawChange					0.01
					(0.00)
N	1138	1068	992	992	992
AIC	2984.88	2807.95	2621.69	2509.92	2510.42
BIC	3125.92	2967.11	2798.08	2803.90	2804.41
$\log L$	−1464.44	−1371.98	−1274.85	−1194.96	−1195.21

Standard errors in parentheses
† Significant at $p < .10$; *$p < .05$; **$p < .01$; ***$p < .001$

Table 6.8 Results of negative binomial regression for count of private activities

	Model 1	Model 2	Model 3	Model 4	Model 5
(Intercept)	0.67***	0.68***	0.62**	0.59**	0.58**
	(0.18)	(0.19)	(0.21)	(0.22)	(0.22)
distance (comdist)	0.12†	0.12†	0.12†	0.13†	0.13†
	(0.06)	(0.06)	(0.07)	(0.07)	(0.07)
majority	0.47***	0.46***	0.48***	0.48***	0.47***
	(0.12)	(0.12)	(0.12)	(0.13)	(0.13)
distance · majority	−0.06	−0.08	−0.08	−0.08	−0.09
	(0.07)	(0.07)	(0.08)	(0.10)	(0.10)
C_m	−4.13***	−4.17***	−4.19***	−4.36***	−4.33***
	(0.55)	(0.57)	(0.61)	(0.62)	(0.62)
C_D	0.97***	0.94***	0.95***	0.94***	0.93***
	(0.16)	(0.16)	(0.17)	(0.18)	(0.18)
nPar		0.00	0.00	0.00	
		(0.00)	(0.00)	(0.00)	
nLawChange					0.00
					(0.00)
access			0.08	0.36*	0.36*
			(0.14)	(0.16)	(0.16)
Business				0.07	0.07
				(0.11)	(0.11)
Prof. Soc.				0.12	0.11
				(0.42)	(0.42)
Public IG				0.07	0.06
				(0.09)	(0.10)
Ministries				−0.04	−0.05
				(0.14)	(0.13)
Parties				−0.07	−0.07
				(0.10)	(0.10)
States				−0.37**	−0.38**
				(0.13)	(0.13)
N	1113	1043	973	973	973
AIC	2770.87	2596.77	2414.82	2412.51	2411.92
BIC	2911.29	2755.17	2590.52	2705.33	2704.74
log L	−1357.44	−1266.39	−1171.41	−1146.25	−1145.96

Standard errors in parentheses
†Significant at $p < .10$; *$p < .05$; **$p < .01$; ***$p < .001$

REFERENCES

Bates, D., Maechler, M., Bolker, B., & Walker, S. (2013). *lme4: Linear Mixed-Effects Models Using Eigen and S4*. R package version 1.0-5. http://CRAN.R-project.org/package=lme4.

Binderkrantz, A. (2008). Different Groups, Different Strategies: How Interest Groups Pursue Their Political Ambitions. *Scandinavian Political Studies, 31*(2), 173–200.

Dür, A. (2008). Measuring Interest Group Influence in the EU. A Note on Methodology. *European Union Politics, 9*(4), 559–576.

Dür, A., & de Bièvre, D. (2007). The Question of Interest Group Influence. *Journal of Public Policy, 27*(01), 1–12.

Gais, T. L., & Walker, J. L. (2001). Pathways to Influence in American Politics. In J. L. Walker (Ed.), *Mobilizing Interest Groups in America*, Chapter 6 (pp. 103–121). Ann Arbor: University of Michigan Press.

Gelman, A., & Hill, J. (2007). *Data Analysis Using Regression and Multilevel/Hierarchical Models*. Cambridge: Cambridge University Press.

Hall, R. L., & Deardorff, A. V. (2006). Lobbying as Legislative Subsidy. *American Political Science Review, 100*(1), 69–84.

Hox, J. J. (2010). *Multilevel Analysis* (2nd ed.). London: Routledge.

Long, J. S. (1997). *Regression Models for Categorical and Limited Dependent Variables*. Thousand Oaks: Sage.

Luce, R. D., & Raiffa, H. (1957). *Games and Decisions: Introduction and Critical Survey*. Mineola: Dover.

Mahoney, C. (2004). The Power of Institutions: State and Interest Group Activity in the European Union. *European Union Politics, 5*(4), 441–466.

Samii, C. (2016). Causal Empiricism in Quantitative Research. *Journal of Politics 78*(3), 941–955. https://doi.org/10.1086/686690.WOS: 000378728000032.

Simon, H. A. (1953). Notes on the Observation and Measurement of Political Power. *Journal of Politics, 15*(4), 500–516.

Train, K. E. (2009). *Discrete Choice Methods with Simulation*. Cambridge: Cambridge University Press.

Wooldridge, J. M. (2002). *Econometric Analysis of Cross Section and Panel Data*. Cambridge: The MIT Press.

Young, I. T. (1977). Proof Without Prejudice: Use of the Kolmogorov-Smirnov Test for the Analysis of Histograms from Flow Systems and Other Sources. *Journal of Histochemistry & Cytochemistry, 25*(7), 935–941.

CHAPTER 7

Conclusions

Abstract In this chapter, I revisit the research question. I discuss the empirical and theoretical approach of the book. I conclude that interest group activities can be classified into equivalence classes. The choice of strategies can be explained by the distance to the constraining political actor (private communication) or the expected policy outcome (public communication). These effects are still valid when controlling for group type, which has been one of the main explanatory factors for strategy choice. I demonstrate that it is necessary to think beyond group type.

Keywords Lobbying · Decision-making · Communication strategy

Lobbying is the backbone of representative democracy. While elections are the main vehicle for selecting politicians and holding them accountable, lobbying is the principal means to influence public policies between elections. In fact, lobbying is so deeply intertwined with politics that understanding policy outcomes is rarely possible without taking lobbying into account. While lobbying is often seen in a very negative light in public discourses, it is often seen more benevolently in scientific discourses.

One of the main aspects of lobbying is the provision of information to policymakers. Interest groups are better informed than politicians on specific effects and aspects of public policies. It seems natural that politicians would rely on this expertise to draft better legislation (Esterling 2004; Truman 1971). However, one problem arises: By definition, interest groups are self-interested political actors, pushing their own agenda. This leads to the important question when will interest groups share their information

© The Author(s) 2019   143
S. Koehler, *Lobbying, Political Uncertainty and Policy Outcomes,*
https://doi.org/10.1007/978-3-319-97055-4_7

truthfully with policymakers and when will they try to misrepresent the information? The answer to this question depends crucially on the communication strategy of interest groups.

This book set out to improve our understanding of interest groups' choice of lobbying strategies, which are understood as the way in which an interest group communicates with politicians. It is crucial to take the political context of lobbying into account. Any attempt to try and understand lobbying strategies without reference to the political process in which lobbying is embedded will at best provide a partial understanding of interest group strategy choice.

While my approach has a strong micro foundation, it moves beyond the numerous attempts which try to explain interest group strategy choice by individual group characteristics. These often fail to take the context of actions into account. But interest groups do not always use the same strategies. I have shown that some of the strategies are structurally equivalent once one conceives their intended purpose as the reference.

Many of the studies who engage with these questions are focused on the US Congress (e.g. Kollman 1997; Baumgartner et al. 2009). While the research on Congress helped us in many ways to enhance our knowledge of lobbying processes, it is hard to generalize to other political systems due to the many peculiarities of the political process. Similarly, theoretical work is largely institution free in the sense that generic communication processes are modeled (e.g. Crawford and Sobel 1982; Potters and van Winden 1992).

The political process is a prime example of a situation of decision-making under uncertainty. Legislators face uncertainty when deciding whether to adopt or reject a bill. This *fundamental* uncertainty results from the difference between the policy that is adopted and the outcome of the policy. The demand for lobbying is created by politicians' dependence on information which is essential to make a better decision. Interest groups, on the other hand, face *process uncertainty* generated by the bargaining environment. The logic of political decision-making in a parliamentary system with proportional representation is very different from the US context.

Adapting the existing models to the setting of a parliamentary system was the first challenge to be tackled in the book. Toward this end, I developed a game theoretic model of lobbying which takes into account the institutional setup in a parliamentary system such as Germany. The model illuminates the differences in the bargaining environment and how they structure incentives for lobbying. The activities of interest groups were modeled by combining a signaling model of lobbying with a veto bargaining model. The model allowed me to open the black box of the single decision-maker in standard models of informational lobbying (e.g. Potters and van Winden 1992).

Similar to standard models, the assumption is that interest groups are better informed about the fundamental causal mechanisms. Interest groups possess valuable knowledge about how policies translate into policy outcomes, they face no *fundamental* uncertainty. However, as political decision-making is a bargaining situation in which the outcome is the result of a legislative process, the outcome of this process is usually not certain. The interest group, therefore, faces a serious amount of uncertainty about the potential outcome of its lobbying strategy. I have called this uncertainty *process uncertainty*. A couple of factors influences the degree of *process uncertainty*. One of the most important is the level of fundamental uncertainty. As such, the two types of uncertainty are closely connected. This is where my model deviates significantly from standard models of lobbying.

Due to process uncertainty, interest groups lack full knowledge of the exact outcome of their lobbying strategy ex-ante. This is often overlooked in analyses of lobbying but has strong implications. The exchange relations inherent to the lobbying situation are characterized by a double uncertainty. The higher the fundamental uncertainty, the higher the demand for information, i.e. lobbying. At the same time, interest groups become less able to predict the policy outcome. In order to understand interest group strategy choices, it is vital to look at what happens after the influence attempt from the interest group's point of view.

The choice of lobbying strategy encompasses two interrelated decisions: The first is to become active or not, i.e. when do interest groups mobilize? The second is—conditional on mobilization—which type of communication does an interest group choose? Choosing a strategy implies to opt for a target and a mode of communication. The mode concerns private *vs.* public communication activities. Potential targets of the communication acts are the relevant decision-makers, which can be either the chambers in a bicameral legislature or the coalition partners in government in the unicameral case.

For understanding process uncertainty and how it affects interest groups' incentives to provide information, it is essential to see the world from an interest groups perspective. It is central to identify the likely outcomes which result from different lobbying strategies in order to understand what drives interest groups' activities regarding a specific bill. Working toward a prediction of interest group behavior, I have turned the problem around and asked the question what is the expected policy outcome conditional on the interest groups strategy. Interest groups base their choice of communication strategy on the *expected* outcome and the expected variation in the outcome. The latter part can be considerable due to the risk aversion of interest groups.

The bargaining stage considerably changes the strategic incentives of the interest group compared to a situation where the interest group signals to one decision-maker. A second dimension is added to the problem as by choosing a communication strategy the interest group can alter the range and variation of policy outcomes but not necessarily the expected policy outcome. Thus, the interest group faces two—possibly opposing—incentives: (a) While a communication strategy may shift the expected outcome in a more favorable direction, and (b) this may come at the expense of higher variability of outcomes. This trade-off is the result of the interdependence of the decision of the political actors.

My model implies that interest group is less powerful than is widely believed. Interest groups which are against a change of the Status Quo will lobby not to prevent a change, which is impossible, but to reduce the variation of outcomes. They are more likely to use a public communication strategy compared to groups who are in favor of a change of policy. The latter are more likely to use a strategy which involves private messages.

Interest groups mainly affect the variation of outcomes by changing the information structure in the political system. However, they can only modestly affect expected policy outcomes. This potentially opens a new direction for empirical research as it shows that many scholars may have been looking for effects in the wrong places.

The interest group's ability to affect political bargaining varies considerably with the degree of fundamental uncertainty. The influence of interest groups increases with the level of fundamental uncertainty as for higher degrees of uncertainty a change of the Status Quo without interference by the interest group is less likely.

The constraining effect of the compromising actor is also stronger. Ceteris paribus, a higher level of process uncertainty implies lower fundamental uncertainty as the relative size of the two types of uncertainty matters.

Another factor is the risk aversion of interest groups which creates trade-offs between expectation and variance of policies. In case of low uncertainty, the utility loss from the variation of policy outcomes is indeed greater than the gains from changes in expected policies.

Decision-making institutions thus play a major role in the determination of policy outcomes. They feed back on interest group strategies as the groups need to anticipate the likely results of their activities. By highlighting preference constellations, I move beyond approaches who only explain lobbying by the context of a bill.

I have derived several hypotheses from the model. First, interest groups which prefer a move away from the Status Quo are more likely to use private messages, while interest groups opposing a change of the Status Quo are using more public strategies. I have operationalized this by the distance of the interest group to the center of the bargaining range. There is strong empirical support for this hypothesis, although the effect is reversed for opposition bills. This suggests that differences in interest group behavior between opposition and majority bills deserve more attention.

Likewise, as predicted by the theoretical model, pro-change groups are much more constrained in their lobbying activities and on average less active than anti-change groups. Other, predictions are that higher costs imply less activity and that higher fundamental uncertainty increases lobbying activity. The data strongly support both claims.

I have used both standard logistic regressions and cross-classified hierarchical models to estimate the effects. While the latter account for likely correlations in the error structure of the data, the results are identical. I have shown in several additional analyses with alternative specifications and alternative operationalizations of the dependent variable that the results are robust.

My endeavor has demonstrated the potential to derive better predictions of the lobbying strategies of interest groups if one take political processes and preferences into account. The perils of oversimplifying by clustering interest groups into types or by ignoring the political process cannot be overstated. More work should be dedicated to disentangle different aspects of interest group's utility calculus to better understand their actions. One aspect that should receive more attention in the future is how interest group competition affects the choice of strategies.

If uncertainty is not too high, interest groups affect mainly the variance rather than the mean of the policy outcome. This is an important feature of the model and may explain many of the non-findings in empirical research. If the expected outcome is (almost) the same with or without lobbying, a mean centered idea of causality will never lead to strong empirical insights.

Once a causal effect is defined in terms of the difference in variances as Braumoeller (2006) does, one may be able to find empirical evidence on the effectiveness of interest groups. My findings call for a new approach to empirical analyses of the determinants and effects of interest group activities. Current approaches center mainly on policy outcomes. A more nuanced approach that analyzes the variance of political outcomes as a function of political uncertainty may be the more fruitful approach. A major advantage is that political uncertainty varies with the institutional setup of a polity and thus my model opens new possibilities for comparative research designs. The model strongly suggests to direct the effort more into this direction.

The formal model has several normative implications, which are in line with the empirical findings. First, similar to other models of information transmission, very extreme pro-change interest groups are effectively taken out of the equation. This can be seen in the distribution of ideal points. There are no ideal points more extreme than the political decision-makers on the pro-change side. Extreme pro-change groups are unable to truthfully communicate with decision-makers. The presence of a moderate second chamber (or coalition partner in a unicameral setting) creates this constraint for both public and private communication.

At the same time, compared to a moderate chamber or coalition party, the presence of a relatively progressive second chamber increases the possibility of information transmission by public messages. The moderate chamber is the constraining factor in the political process. Inducing change requires loosening this constraint. More groups will be able to truthfully communicate with this chamber using public compared to private communication. Public communication is therefore not necessarily evidence for Status Quo groups trying to preserve public policies.

Second, in the cases where truthful communication is possible, the presence of lobbying enhances the welfare of both the interest group and the decision-makers. To the degree that decision-makers' preferences reflect preferences of their constituencies, this also enhances the welfare of society as a whole. Biased information can be beneficial.

Lastly, the model implies that there may be a new rationale for public communication by Status Quo groups. It is often claimed that public communication has three goals: Signaling to constituencies, mobilization and conflict expansion (Kollman 1998).

In parliamentary systems, these goals are important as well. However, given that interest groups are unable to prevent change once political decision-makers agree to change public policies, risk aversion may add another dimension: The reduction of variation in public policies. Ex-ante, interest groups face uncertainty about the exact outcome of the political process. Successful public communication will not affect the (ex-ante) expected outcome, but will reduce variability of outcomes. Risk averse interest groups do value this, as it makes the process more predictable.

The predictions of the theoretical model are less accurate for opposition bills. Interest groups which are in favor of a change of the Status Quo are more likely to send public messages in case of opposition bills. Given that opposition bills have little chances of being accepted in the German parliamentary democracy, this seems to imply a difference in the strategic calculus of interest groups. This case is not directly covered by the model and would require further analysis. A possible explanation is that the messages are a signal to the group's constituency helping to ensure group maintenance.

Future work should also take the claim serious that some of what lobbyists do affects the variance rather than the (expected) outcome of policies. A main challenge will be the operationalization and measurement of these concepts and to develop a clear empirical strategy to test for the effect.

REFERENCES

Baumgartner, F. R., Berry, J. M., Hojnacki, M., Kimball, D. C., & Lech, B. L. (2009). *Lobbying and Policy Change: Who Wins, Who Loses, and Why*. Chicago: University of Chicago Press.

Braumoeller, B. F. (2006). Explaining Variance; Or Stuck in a Moment We Can't Get Out Of. *Political Analysis, 14*(1), 268–290.

Crawford, V. P., & Sobel, J. (1982). Strategic Information Transmission. *Econometrica, 50*(6), 1431–1451.

Esterling, K. M. (2004). *The Political Economy of Expertise: Information and Efficiency in American National Politics*. Ann Arbor: The University of Michigan Press.

Kollman, K. (1997). Inviting Friends to Lobby: Interest Groups, Ideological Bias, and Congressional Committees. *American Journal of Political Science, 41*(2), 519–544.

Kollman, K. (1998). *Outside Lobbying*. Princeton, NJ: Princeton University Press.

Potters, J., & van Winden, F. (1992). Lobbying and Asymmetric Information. *Public Choice, 74*(3), 269–292.

Truman, D. B. (1971). *The Governmental Process. Political Interests and Public Opinion* (2nd ed.). New York: Knopf.

Bibliography

Ainsworth, S. H. (1997). The Role of Legislators in the Determination of Interest Group Influence. *Legislative Studies Quarterly, 22*(4), 517–533.

Ansolabehere, S., Snyder, J. M., & Ting, M. M. (2003). Bargaining in Bicameral Legislatures: When and Why Does Malapportionment Matter? *The American Political Science Review, 97*(3), 471–481.

Austen-Smith, D. (1993). Information and Influence: Lobbying for Agendas and Votes. *American Journal of Political Science, 37*(3), 799–833.

Austen-Smith, D. (1997). Interest Groups: Money, Information and Influence. In D. C. Mueller (Ed.), *Perspectives on Public Choice* (pp. 296–322). Cambridge: Cambridge University Press.

Austen-Smith, D. (1998). Allocating Access for Information and Contributions. *Journal of Law Economics & Organization, 14*(2), 277–303.

Austen-Smith, D., & Banks, J. S. (2000). Cheap Talk and Burned Money. *Journal of Economic Theory, 91*(1), 1–16.

Austen-Smith, D., & Banks, J. S. (2002). Costly Signaling and Cheap Talk in Models of Political Influence. *European Journal of Political Economy, 18*(2), 263–280.

Austen-Smith, D., & Wright, J. R. (1992). Competitive Lobbying for a Legislator's Vote. *Social Choice and Welfare, 9*(3), 229–257.

Austen-Smith, D., & Wright, J. R. (1994). Counteractive Lobbying. *American Journal of Political Science, 38*(1), 25–44.

Austen-Smith, D., & Wright, J. R. (1996). Theory and Evidence for Counteractive Lobbying. *American Journal of Political Science, 40*(2), 543–564.

© The Editor(s) (if applicable) and The Author(s) 2019
S. Koehler, *Lobbying, Political Uncertainty and Policy Outcomes*,
https://doi.org/10.1007/978-3-319-97055-4

Baron, D. P. (1989). A Noncooperative Theory of Legislative Coalitions. *American Journal of Political Science, 33*(4), 1048–1084.

Baron, D. P. (2006). Competitive Lobbying and Supermajorities in a Majority-Rule Institution. *Scandinavian Journal of Economics, 108*(4), 607–642.

Baron, D. P., & Ferejohn, J. A. (1989). Bargaining in Legislatures. *American Political Science Review, 83*(4), 1181–1206.

Bates, D., Maechler, M., Bolker, B., & Walker. S. (2013). *lme4: Linear Mixed-Effects Models Using Eigen and S4.* R Package Version 1.0-5. http://CRAN.R-project.org/package=lme4.

Battaglini, M. (2002). Multiple Referrals and Multidimensional Cheap Talk. *Econometrica, 70*(4), 1379–1401.

Baumgartner, F. R. (2007). Eu Lobbying: A View from the US. *Journal of European Public Policy, 14*(3), 482–488.

Baumgartner, F. R., Berry, J. M., Hojnacki, M., Kimball, D. C., & Lech, B. L. (2009). *Lobbying and Policy Change: Who Wins, Who Loses, and Why.* Chicago: University of Chicago Press.

Baumgartner, F. R., & Jones, B. D. (1993). *Agendas and Instability in American Politics.* Chicago: University of Chicago Press.

Baumgartner, F. R., & Leech, B. L. (1998). *Basic Interests.* Princeton, NJ: Princeton University Press.

Becker, G. S. (1983). A Theory of Competition Among Pressure Groups for Political Influence. *The Quarterly Journal of Economics, 98*(3), 371–400.

Becker, G. S. (1985). Public Policies, Pressure Groups, and Dead Weight Costs. *Journal of Public Economics, 28*(3), 329–347.

Bentley, A. F. (1908). *The Process of Government.* Chicago: University of Chicago Press.

Bernhagen, P. (2008a). Business and International Environmental Agreements: Domestic Sources of Participation and Compliance by Advanced Industrialized Democracies. *Global Environmental Politics, 8*(1), 78–110.

Bernhagen, P. (2008b). *The Political Power of Business: Structure and Information in Public Policymaking.* London: Chapman & Hall.

Bernhagen, P., & Bräuninger, T. (2005). Structural Power and Public Policy: A Signaling Model of Business Lobbying in Democratic Capitalism. *Political Sudies, 53*(1), 43–64.

Bernheim, B. D., & Whinston, M. D. (1986a). Common Agency. *Econometrica, 54*(4), 923–942.

Bernheim, B. D., & Whinston, M. D. (1986b). Menu Auctions, Resource-Allocation, and Economic Influence. *Quarterly Journal of Economics, 101*(1), 1–31.

Beyers, J. (2004). Voice and Access: Political Practices of European Interest Associations. *European Union Politics, 5*(2), 211–240.

Beyers, J., & Kerremans, B. (2007). The Press Coverage of Trade Issues: A Comparative Analysis of Public Agenda-Setting and Trade Politics. *Journal of European Public Policy, 14*(2), 269–292.

Binderkrantz, A. (2003). Strategies of Influence: How Interest Organizations React to Changes in Parliamentary Influence and Activity. *Scandinavian Political Studies, 26*(4), 287–306.

Binderkrantz, A. (2005). Interest Group Strategies: Navigating Between Privileged Access and Politics of Pressure. *Political Studies, 53*, 694–715.

Binderkrantz, A. (2008). Different Groups, Different Strategies: How Interest Groups Pursue Their Political Ambitions. *Scandinavian Political Studies, 31*(2), 173–200.

Binmore, K., Rubinstein, A., & Wolinsky, A. (1986). The Nash Bargaining Solution in Economic Modeling. *Rand Journal of Economics, 17*(2), 176–188.

Birchler, U., & Bütler, M. (2007). *Information Economics*. London: Routledge.

Black, D. (1958). *The Theory of Committees and Elections*. London: Cambridge University Press.

Bouwen, P. (2002). Corporate Lobbying in the European Union: The Logic of Access. *Journal of European Public Policy, 9*(3), 365–390.

Bouwen, P. (2004a). Exchanging Access Goods for Access: A Comparative Study of Business Lobbying in the European Union Institutions. *European Journal of Political Research, 43*(3), 337–369.

Bouwen, P. (2004b). The Logic of Access to the European Parliament: Business Lobbying in the Committee on Economic and Monetary Affairs. *Journal of Common Market Studies, 42*(3), 473–496.

Box-Steffensmeier, J. M., Brady, H. E., & Collier, D. (Eds.). (2008). *The Oxford Handbook of Political Methodology*. Oxford: Oxford University Press.

Braumoeller, B. F. (2006). Explaining Variance: Or Stuck in a Moment We Can't Get Out Of. *Political Analysis, 14*(1), 268–290.

Bräuninger, T., & König, T. (2004). Senden und Empfangen: Informationstransfer in Politiknetzwerken als Vermittlung von Verhandlungsvorschlägen. In C. H. Henning & C. Melbeck (Eds.), *Interdisziplinäre Sozialforschung. Theorie und empirische Anwendungen* (pp. 205–224). Frankfurt: Campus Verlag.

Cameron, C. (2000). *Veto Bargaining*. Cambridge: Cambridge University Press.

Cameron, C., & McCarty, N. (2004). Models of Vetoes and Veto Bargaining. *Annual Review of Political Science, 7*(1), 409–435.

Carpenter, D. P., Esterling, K. M., & Lazer, D. M. J. (2004). Friends, Brokers, and Transitivity: Who Informs Whom in Washington Politics? *The Journal of Politics, 66*(1), 224–246.

Clinton, J., Jackman, S., & Rivers, D. (2004). The Statistical Analysis of Roll Call Data. *American Political Science Review, 98*(2), 355–370.

Coen, D. (1997). The Evolution of the Large Firm as a Political Actor in the European Union. *Journal of European Public Policy, 4*(1), 91–108.

Coen, D. (1998). The European Business Interest and the Nation State: Large-Firm Lobbying in the European Union and Member States. *Journal of Public Policy*, *18*(1), 75–100.

Coen, D. (2007). Empirical and Theoretical Studies in EU Lobbying. *Journal of European Public Policy*, *14*(3), 333–345.

Cotton, C. (2012). Pay-to-Play Politics: Informational Lobbying and Contribution Limits When Money Buys Access. *Journal of Public Economics*, *96*(3/4), 369–386.

Cotton, C. (2016). Competing for Attention: Lobbying Time-Constrained Politicians. *Journal of Public Economic Theory*, *18*(4), 642–665.

Crawford, V. P., & Sobel, J. (1982). Strategic Information Transmission. *Econometrica*, *50*(6), 1431–1451.

Crombez, C. (1996). Legislative Procedures in the European Community. *British Journal of Political Science*, *26*, 199–228.

Crombez, C. (2000). Institutional Reform and Co-decision in the European Union. *Constitutional Political Economy*, *11*(1), 41–57.

Crombez, C. (2001). The Treaty of Amsterdam and the Codecision Procedure in the European Union. In G. Schneider & M. Aspinwall (Eds.), *The Rules of Integration. The Institutionalist Approach to European Studies*. Manchester: Manchester University Press.

Crombez, C. (2002). Information, Lobbying and the Legislative Process in the European Union. *European Union Politics*, *3*(1), 7–32.

Crombez, C., Groseclose, T., & Krehbiel, K. (2006). Gatekeeping. *Journal of Politics*, *68*(2), 322–334.

Cutrone, M., & McCarty, N. (2009). Does Bicameralism Matter? In B. R. Weingast & D. A. Wittman (Eds.), *Oxford Handbook of Political Economy, Chapter 10* (pp. 180–195). Oxford: Oxford University Press.

Dixit, A., Grossman, G. M., & Helpman, E. (1997). Common Agency and Co-ordination: General theory and Application to Government Policy Making. *The Journal of Political Economy*, *105*(4), 752–769.

Dür, A. (2008a). Bringing Economic Interests Back into the Study of EU Trade Policy-Making. *British Journal of Politics & International Relations*, *10*(1), 27–45.

Dür, A. (2008b). Measuring Interest Group Influence in the EU. A Note on Methodology. European Union. *Politics*, *9*(4), 559–576.

Dür, A., & de Bièvre, D. (2007a). Inclusion Without Influence? NGOs in European Trade Policy. *Journal of Public Policy*, *27*(01), 79–101.

Dür, A., & de Bièvre, D. (2007b). The Question of Interest Group Influence. *Journal of Public Policy*, *27*(01), 1–12.

Dür, A., & Mateo, G. (2013). Gaining Access or Going Public? Interest Group Strategies in Five European Countries. *European Journal of Political Research*, *52*(5), 660–686.

Dür, A., & Mateo, G. (2016). *Insiders Versus Outsiders: Interest Group Politics in Multilevel Europe*. Oxford: Oxford University Press.

Dusso, A. (2010). Legislation, Political Context, and Interest Group Behavior. *Political Research Quarterly*, *63*(1), 55–67.

Duverger, M. (1951). *Political Parties: Their Organization and Activity in the Modern State*. New York: Wiley.

Easton, D. (1965a). *Framework for Political Analysis*. Upper Saddle River: Prentice Hall.

Easton, D. (1965b). *A Systems Analysis of Political Life*. New York: Wiley.

Edgeworth, F. Y. (1881). *Mathematical Psychics*. London: Kegan Paul.

Eising, R. (2004). Multilevel Governance and Business Interests in the European Union. *Governance*, *17*(2), 211–245.

Eising, R. (2007a). The Access of Business Interests to EU Institutions: Towards Élite Pluralism? *Journal of European Public Policy*, *14*(3), 384–403.

Eising, R. (2007b). Institutional Context, Organizational Resources and Strategic Choices: Explaining Interest Group Access in the European Union. *European Union Politics*, *8*(3), 329–362.

Esterling, K. M. (2004). *The Political Economy of Expertise: Information and Efficiency in American National Politics*. Ann Arbor: The University of Michigan Press.

Farrell, J., & Gibbons, R. (1989). Cheap Talk with Two Audiences. *The American Economic Review*, *79*(5), 1214–1223.

Fishburn, P. C., & Rubinstein, A. (1982). Time Preference. *International Economic Review*, *23*(3), 677–694.

Gais, T. L., & Walker, J. L. (2001). Pathways to Influence in American Politics. In J. L. Walker (Ed.), *Mobilizing Interest Groups in America, Chapter 6* (pp. 103–121). Ann Arbor: University of Michigan Press.

Garrett, G. (1995). From the Luxembourg Compromise to Codecision: Decision Making in the European Union. *Electoral Studies*, *14*(3), 289–308.

Gelman, A., & Hill, J. (2007). *Data Analysis Using Regression and Multilevel/Hierarchical Models*. Cambridge: Cambridge University Press.

Gilligan, T. W., & Krehbiel, K. (1987). Collective Decisionmaking and Standing Committees: An Informational Rationale for Restrictive Amendment Procedures. *Journal of Law and Economic Organization*, *3*(2), 287–335.

Goldstein, K. M. (1999). *Interest Groups, Lobbying, and Participation in America*. Cambridge: Cambridge University Press.

Greenwood, J. (2007). *Interest Representation in the European Union* (2nd ed.). Houndmills, Basingstoke: Palgrave Macmillan.

Grossman, G. M., & Helpman, E. (1994). Protection for Sale. *American Economic Review*, *84*(4), 833–850.

Grossman, G. M., & Helpman, E. (1996). Electoral Competition and Special Interest Politics. *The Review of Economic Studies*, *63*(2), 265–286.

Grossman, G. M., & Helpman, E. (2001). *Special Interest Politics*. Cambridge, MA: The MIT Press.

Grossman, S. J., & Perry, M. (1986). Sequential Bargaining Under Asymmetric Information. *Journal of Economic Theory, 39*(1), 120–154.

Gullberg, A. T. (2008a). Lobbying Friend and Foes in Climate Policy: The Case of Business and Environmental Interest Groups in the European Union. *Energy Policy, 36*(8), 2964–2972.

Gullberg, A. T. (2008b). Rational Lobbying and EU Climate Policy. *International Environmental Agreements: Politics, Law and Economics, 8*(2), 161–178.

Hall, R. L., & Deardorff, A. V. (2006). Lobbying as Legislative Subsidy. *American Political Science Review, 100*(1), 69–84.

Hall, R. L., & Reynolds, M. E. (2012). Targeted Issue Advertising and Legislative Strategy: The Inside Ends of Outside Lobbying. *American Journal of Political Science, 74*(3), 888–902.

Hall, R. L., & Wayman, F. W. (1990). Buying Time: Moneyed Interests and the Mobilization of Bias in Congressional Committees. *American Political Science Review, 84*(3), 797–820.

Hedström, P. (2008). Studying Mechanisms to Strengthen Causal Inferences in Quantitative Research. In J. Box-Steffensmeier, H. E. Brady, & D. Collier (Eds.), *Oxford Handbook of Political Methodology* (pp. 319–335). Oxford: Oxford University Press.

Heinz, J. P., Laumann, E. O., Nelson, R. L., & Salisbury, R. H. (1993). *The Hollow Core*. Cambridge: Harvard University Press.

Heller, W. B. (2007). Divided Politics: Bicameralism, Parties, and Policy in Democratic Legislatures. *Annual Review of Political Science, 10*(1), 245–269.

Hillman, A. L., & Riley, J. G. (1989). Politically Contestable Rents and Transfers. *Economics and Politics, 1*(1), 17–39.

Hinich, M. J., & Munger, M. C. (1997). *Analytical Politics*. Cambridge: Cambridge University Press.

Hirshleifer, J., & Riley, J. G. (1991). *The Analytics of Uncertainty and Information*. Cambridge: Cambridge University Press.

Hojnacki, M., & Kimball, D. C. (1998). Organized Interests and the Decision of Whom to Lobby in Congress. *The American Political Science Review, 92*(4), 775–790.

Hojnacki, M., & Kimball, D. C. (1999). The Who and How of Organizations' Lobbying Strategies in Committee. *The Journal of Politics, 61*(4), 999–1024.

Holyoke, T. T. (2003). Choosing Battlegrounds: Interest Group Lobbying Across Multiple Venues. *Political Research Quarterly, 56*(3), 325–336.

Hox, J. J. (2010). *Multilevel Analysis* (2nd ed.). London: Routledge.

Huber, J. D., & McCarty, N. (2001). Cabinet Decision Rules and Political Uncertainty in Parliamentary Bargaining. *The American Political Science Review, 95*(2), 345–360.

Huber, J. D., & Shipan, C. R. (2002). *Deliberate Discretion? The Institutional Foundations of Bureaucratic Autonomy*. Cambridge: Cambridge University Press.

Ismayr, W. (2008a). Gesetzgebung im politischen System Deutschlands. In W. Ismayr (Ed.), *Gesetzgebung in Westeuropa: EU-Staaten und Europäische Union* (pp. 383–430). Wiesbaden: VS Verlag für Sozialwissenschaften.

Ismayr, W. (2008b). *Gesetzgebung in Westeuropa: EU-Staaten und Europäische Union*. Wiesbaden: VS - Verlag für Sozialwissenschaften.

Ismayr, W. (2012). *Der Deutsche Bundestag* (3rd ed.). Wiesbaden: Springer VS.

Jackman, S. (2008). Measurement. In J. Box-Steffensmeier, H. E. Brady, & D. Collier (Eds.), *Oxford Handbook of Political Methodology* (pp. 119–151). Oxford: Oxford University Press.

Johnson, V. E., & Albert, J. H. (1999). *Ordinal Data Modeling*. Heidelberg and New York: Springer.

Kalai, E., & Smorodinsky, M. (1975). Other Solutions to Nash's Bargaining Problem. *Econometrica*, *43*(3), 513–518.

Kleinfeld, R., Zimmer, A., & Willems, U. (Eds.). (2007). *Lobbying-Strukturen. Akteure. Strategien*. Wiesbaden: Springer.

Knoke, D. (1990). *Political Networks: The Structural Perspective*. New York: Cambridge University Press.

Knoke, D., & Burleigh, F. (1989). Collective Action in National Policy Domains: Constraints, Cleavages, and Policy Outcomes. *Research in Political Sociology*, *4*, 187–208.

Knoke, D., & Kaufmann, N. J. (1992). Social Organization of the United States National Labor Policy Domain [Computer File]. Minneapolis, MN: David Knoke, University of Minnesota, Dept. of Sociology [producer]; Ann Arbor, MI: Inter-university Consortium for Political and Social Research [distributor]. http://dx.doi.org/10.3886/ICPSR09802.

Knoke, D., & Pappi, F. U. (1991). Organizational Action Sets in the U.S. and German Labor Policy Domains. *American Sociological Review*, *56*(4), 509–523.

Knoke, D., Pappi, F. U., Broadbent, J., & Tsujinaka, Y. (1996). *Comparing Policy Networks*. Cambridge: Cambridge University Press.

Koessler, F. (2008). Lobbying with Two Audiences: Public vs Private Certification. *Mathematical Social Sciences*, *55*(3), 305–314.

Köhler, S. (2014). *Political Uncertainty and Interest Group Communication Strategies*. Dissertation, University of Mannheim.

Kollman, K. (1997). Inviting Friends to Lobby: Interest Groups, Ideological Bias, and Congressional Committees. *American Journal of Political Science*, *41*(2), 519–544.

Kollman, K. (1998). *Outside Lobbying*. Princeton, NJ: Princeton University Press.

König, T. (1992). *Entscheidungen im Politiknetzwerk*. Wiesbaden: Deutscher Universitäts Verlag.

König, T., & Bräuninger, T. (1998). The Formation of Policy Networks. *Journal of Theoretical Politics*, *10*(4), 445–471.

König, T., & Pappi, F. U. (1989). *Politikfeld Arbeit-Codebuch der Deutschen Teilstudie*. Kiel, Germany: Institut für Soziologie, Christian-Albrechts-University.

Krehbiel, K. (1998). *Pivotal Politics: A Theory of U.S. Lawmaking*. Chicago: University of Chicago Press.

Kreps, D. M. (1990). *A Course in Microeconomic Theory*. Herfortshire: Harvester Wheatsheaf.

Krishna, V., & Morgan, J. (2001). A Model of Expertise. *The Quarterly Journal of Economics*, *116*(2), 747–775.

Langbein, L. I. (1986). Money and Access: Some Empirical Evidence. *The Journal of Politics*, *48*(4), 1052–1062.

Lasswell, H. D. (1950). *Power and Society: A Framework for Political Inquiry*. New Haven: Yale University Press.

Laumann, E. O., & Knoke, D. (1987). *The Organizational State: A Perspective on National Energy and Health Domains*. Madison: University of Wisconsin Press.

Leblond, P. (2008). The Fog of Integration: Reassessing the Role of Economic Interests in European Integration. *British Journal of Politics & International Relations*, *10*(1), 9–26.

Leech, B. L., Baumgartner, F. R., Pira, T. M. L., & Semanko, N. A. (2005). Drawing Lobbyists to Washington: Government Activity and the Demand for Advocacy. *Political Research Quarterly*, *58*(1), 19–30.

Lijphart, A. (1999). *Patterns of Democracy: Government Forms and Performance in Thirty-Six Countries*. Yale: Yale University Press.

Lindberg, L. N. (1963). *The Political Dynamics of European Economic Integration*. Stanford: Stanford University Press.

Lohmann, S. (1995). Information, Access, and Contributions: A Signalling Model of Lobbying. *Public Choice*, *85*(3–4), 267–284.

Lohmann, S. (2003). Representative Government and Special Interest Politics (We Have Met the Enemy and He Is Us). *Journal of Theoretical Politics*, *15*(3), 299–319.

Long, J. S. (1997). *Regression Models for Categorical and Limited Dependent Variables*. Thousand Oaks: Sage.

Lowery, D., & Gray, V. (2004). Bias in the Heavenly Chorus Interests in Society and Before Government. *Journal of Theoretical Politics*, *16*(1), 5–29.

Lowery, D., Poppelaars, C., & Berkhout, J. (2008). The European Union Interest System in Comparative Perspective: A Bridge Too Far? *West European Politics*, *31*(6), 1231.

Lowi, W. J. (1972). Four Systems of Policy. Politics and Choice. *Public Administration Review*, *32*(4), 298–310.

Luce, R. D., & Raiffa, H. (1957). *Games and Decisions: Introduction and Critical Survey*. Mineola: Dover.

Macho-Stadler, I., & Pérez-Castrillo, D. (2001). *An Introduction to the Economics of Information: Incentives and Contracts* (2nd ed.). Oxford: Oxford University Press.

Mahoney, C. (2004). The Power of Institutions: State and Interest Group Activity in the European Union. *European Union Politics, 5*(4), 441–466.

Mahoney, C. (2007a). Lobbying Success in the United States and the European Union. *Journal of Public Policy, 27*(1), 35–56.

Mahoney, C. (2007b). Networking vs. Allying: The Decision of Interest Groups to Join Coalitions in the US and the EU. *Journal of European Public Policy, 14*(3), 366–383.

Mahoney, C. (2008). *Brussels Versus the Beltway: Advocacy in the United States and the European Union*. Washington, DC: Georgetown University Press.

Mahoney, C., & Baumgartner, F. (2008). Converging Perspectives on Interest Group Research in Europe and America. *West European Politics, 31*(6), 1253.

Manley, J. F. (1983). Neo-pluralism: A Class Analysis of Pluralism I and Pluralism II. *The American Political Science Review, 77*(2), 368–383.

Martin, L. W., & Vanberg, G. (2004). Policing the Bargain: Coalition Government and Parliamentary Scrutiny. *American Journal of Political Science, 48*(1), 13–27.

Martin, L. W., & Vanberg, G. (2005). Coalition Policy Making and Legislative Review. *The American Political Science Review, 99*(1), 93–106.

Matthews, S. A. (1989). Veto Threats: Rhetoric in a Bargaining Game. *The Quarterly Journal of Economics, 104*(2), 347–369.

McCarty, N. (2000). Proposal Rights, Veto Rights, and Political Bargaining. *American Journal of Political Science, 44*(3), 506–522.

McCubbins, M., & Schwartz, T. (1984). Congressional Oversight Overlooked: Police Patrols Versus Fire Alarms. *American Journal of Political Science, 28*(1), 16–79.

McFadden, D. (1974). Conditional Logit Analysis of Qualitative Choice Behavior. In P. Zarembka (Ed.), *Frontiers in Econometrics* (pp. 105–142). New York: Academic Press.

McFarland, A. S. (2007). Neopluralism. *Annual Review of Political Science, 10*(1), 45–66.

McKay, A. (2008). A Simple Way of Estimating Interest Group Ideology. *Public Choice, 136*, 69–86.

McKay, A. (2012a). Buying Policy? The Effects of Lobbyists' Resources on Their Policy Success. *Political Research Quarterly, 65*(4), 908–923.

McKay, A. (2012b). Negative Lobbying and Policy Outcomes. *American Politics Research, 40*(1), 116–146.

Michalowitz, I. (2007). What Determines Influence? Assessing Conditions for Decision-Making Influence of Interest Groups in the EU. *Journal of European Public Policy, 14*(1), 132–151.

Milyo, J. (2001). What Do Candidates Maximize (and Why Should Anyone Care)? *Public Choice, 109*(1/2), 119–139.

Morton, A. (1993). Mathematical Models: Questions of Trustworthiness. *The British Journal for the Philosophy of Science, 44*(4), 659–674.

Mueller, D. C. (Ed.). (1997). *Perspectives on Public Choice.* Cambridge: Cambridge University Press.

Muthoo, A. (1999). *Bargaining theory with Applications.* Cambridge: Cambridge University Press.

Naoi, M., & Krauss, E. (2009). Who Lobbies Whom? Special Interest Politics Under Alternative Electoral Systems. *American Journal of Political Science, 53*(4), 874–892.

Nash, J. F. (1950). The Bargaining Solution. *Econometrica, 18*(2), 155–162.

Olson, M. (1965). *The Logic of Collective Action: Public Goods and the Theory of Groups.* Cambridge, MA: Harvard University Press.

Osborne, M. J., & Rubinstein, A. (1990). *Bargaining and Markets.* Emerald: Bingley.

Pappi, F. U., & Henning, C. H. C. A. (1998). Policy Networks: More than a Metaphor? *Journal of Theoretical Politics, 10*(4), 553–575.

Pappi, F. U., König, T., & Knoke, D. (1995). *Entscheidungsprozesse in der arbeits- und Sozialpolitik.* Frankfurt/M.: Campus Verlag.

Peltzman, S. (1976). Toward a More General Theory of Regulation. *Journal of Law and Economics, 19*(2), 211–240.

Poole, K. T., & Rosenthal, H. (1985). A Spatial Model for Legislative Roll Call Analysis. *American Journal of Political Science, 29*(2), 357–384.

Potters, J., & Sloof, R. (1996). Interest Groups: A Survey of Empirical Models That Try to Assess Their Influence. *European Journal of Political Economy, 12*, 403–442.

Potters, J., & van Winden, F. (1990). Modelling Political Pressure as Transmission of Information. *European Journal of Political Economy, 6*(1), 61–88.

Potters, J., & van Winden, F. (1992). Lobbying and Asymmetric Information. *Public Choice, 74*(3), 269–292.

Powell, R. (2002). Bargaining Theory and International Conflict. *Annual Review of Political Science, 5*, 1–30.

Princen, S. (2007). Advocacy Coalitions and the Internationalization of Public Health Policies. *Journal of Public Policy, 27*(1), 13–33.

R Core Team. (2018). *R: A Language and Environment for Statistical Computing.* Vienna, Austria: R Foundation for Statistical Computing. http://www.R-project.org/.

Riker, W. H. (1992). The Justification of Bicameralism. *Revue Internationale de Science Politique [International Political Science Review], 13*(1), 101–116.

Rogers, J. R. (1998). Bicameral Sequence: Theory and State Legislative Evidence. *American Journal of Political Science, 42*(2), 1025–1060.

Romer, T., & Rosenthal, H. (1978). Political Resource Allocation, Controlled Agendas, and the Status Quo. *Public Choice*, *33*(4), 27–43.

Rubinstein, A. (1982). Perfect Equilibrium in a Bargaining Model. *Econometrica*, *50*(1), 97–109.

Rubinstein, A. (1985). A Bargaining Model with Incomplete Information About Time-Preferences. *Econometrica*, *53*(5), 1151–1172.

Rudzio, W. (2011). *Das Politische System der Bundesrepublik Deutschland* (8th ed.). Wiesbaden: VS—Verlag für Sozialwissenschaften.

Russel, M. (2001). What Are Second Chambers for? *Parliamentary Affairs*, *54*, 442–458.

Samii, C. (2016). Causal Empiricism in Quantitative Research. *Journal of Politics*, *78*(3), 941–955. https://doi.org/10.1086/686690.WOS: 000378728000032.

Schattschneider, E. E. (1975). *The Semisovereign People: A Realist's View of Democracy in America*. Hinsdale, Ill.: Dryden Press.

Schelling, T. C. (1960). *The Strategy of Conflict*. Cambridge, MA: Harvard University Press.

Schlozman, K. L., & Tierney, J. T. (1986). *Organized Interests and American Democracy*. New York: Longman Higher Education.

Sebaldt, M. (2007). Lobbying in Deutschland - Begriff und Trends. In R. Kleinfeld, A. Zimmer, & U. Willems (Eds.), *Lobbying - Strukturen. Akteure. Strategien* (pp. 92–94). Wiesbaden: Springer.

Selck, T. J., & Steunenberg, B. (2004). Between Power and Luck: The European Parliament in the EU Legislative Process. *European Union Politics*, *5*(1), 25–46.

Shepsle, K. A., & Bonchek, M. S. (1997). *Analyzing Politics*. New York: W. W. Norton.

Simon, H. A. (1953). Notes on the Observation and Measurement of Political Power. *Journal of Politics*, *15*(4), 500–516.

Sloof, R. (1998). *Game-Theoretic Models of the Political Influence of Interest Groups*. Boston: Kluwer.

Sloof, R., & van Winden, F. (2000). Show Them Your Teeth First! A Game-Theoretic Analysis of Lobbying and Pressure. *Public Choice*, *104*, 81–120.

Smith, M. P. (2008). All Access Points Are Not Created Equal: Explaining the Fate of Diffuse Interests in the EU. *British Journal of Politics & International Relations*, *10*(1), 64–83.

Ståhl, I. (1972). *Bargaining Theory*. Stockholm: Economic Research Institute.

Steinberg, D. A., & Shih, V. C. (2012). Interest Group Influence in Authoritarian States: The Political Determinants of Chinese Exchange Rate Policy. *Comparative Political Studies*, *45*(11), 1405–1434.

Steunenberg, B. (1994). Decision-Making Under Different Institutional Arrangements-Legislation by the European-Community. *Journal of Institutional and Theoretical Economics*, *150*(4), 642–669.

Steunenberg, B. (1996). Agent Discretion, Regulatory Policymaking, and Different Institutional Arrangements. *Public Choice*, *86*(3–4), 309–339.

Steunenberg, B., & Selck, T. J. (2006). Testing Procedural Models of EU Legislative Decision Making. In R. Thomson, F. N. Stokman, C. H. Achen, & T. König (Eds.), *The European Union Decides* (pp. 54–85). Cambridge: Cambridge University Press.

Stigler, G. J. (1971). The Theory of Economic Regulation. *The Bell Journal of Economics and Management Science*, *2*(1), 3–21.

Stokman, F. N., & Bueno de Mesquita, B. (1994). *European Community Decision Making: Models, Applications, and Comparisons*. New Haven: Yale University Press.

Streeck, W. (1983). Between Pluralism and Corporatism: German Business Associations and the State. *Journal of Public Policy*, *3*(3), 265–84.

Thies, M. F. (2001). Keeping Tabs on Partners: The Logic of Delegation in Coalition Governments. *American Journal of Political Science*, *45*(3), 580–598.

Thomas, C. S. (Ed.). (2004). *Research Guide to U.S. and International Interest Groups*. Westport: Praeger.

Thomson, R., Stokman, F. N., Achen, C. H., & König, T. (Eds.). (2006). *The European Union Decides*. Cambridge: Cambridge University Press.

Train, K. E. (2009). *Discrete Choice Methods with Simulation*. Cambridge: Cambridge University Press.

Truman, D. B. (1971). *The Governmental Process. Political Interests and Public Opinion* (2nd ed.). New York: Knopf.

Tsebelis, G. (1991). *Nested Games: Rational Choice in Comparative Politics*. Berkeley: University of California Press.

Tsebelis, G. (2002). *Veto Players: How Political Institutions Work*. Princeton: Princeton University Press.

Tsebelis, G., & Garrett, G. (1996). Agenda Setting Power, Power Indices, and Decision-Making in the European Union. *International Review of Law and Economics*, *16*(3), 345–361.

Tsebelis, G., & Garrett, G. (2001). The Institutional Foundations of Intergovernmentalism and Supranationalism in the European Union. *International Organization*, *55*(2), 357–390.

Tsebelis, G., & Money, J. (1997). *Bicameralism*. Cambridge: Cambridge University Press.

Tsebelis, G., & Yataganas, X. (2002). Veto Players and Decision-Making in the EU After Nice: Policy Stability and Bureaucratic/Judicial Discretion. *Journal of Common Market Studies*, *40*(2), 283–307.

van Winden, F. (2008). Interest Group Behavior and Influence. In C. K. Rowley & F. Schneider (Eds.), *Readings in Public Choice and Constitutional Political Economy*. New York: Springer.

Victor, J. N. (2007). Strategic Lobbying. Demonstrating How Legislative Context Affects Interest Groups' Lobbying Tactics. *American Politics Research, 35*(6), 826–845.

von Beyme, K. (2010). *Das Politische System der Bundesrepublik Deutschland* (11th ed.). Wiesbaden: VS - Verlag für Sozialwissenschaften.

Ward, H. (2004). Pressure Politics: A Game-Theoretical Investigation of Lobbying and the Measurement of Power. *Journal of Theoretical Politics, 16*(1), 31–52.

Wasserman, S., & Faust, K. (1994). *Social Network Analysis.* Cambridge: Cambridge University Press.

Wehrmann, I. (2007). Lobbying in Deutschland - Begriff und Trends. In R. Kleinfeld, A. Zimmer, & U. Willems (Eds.), *Lobbying - Strukturen. Akteure. Strategien* (pp. 36–64). Wiesbaden: Springer (VS).

Wooldridge, J. M. (2002). *Econometric Analysis of Cross Section and Panel Data.* Cambridge: The MIT Press.

Wright, J. R. (1996). *Interest Groups & Congress.* Needham Heights, MA: Allyn & Bacon.

Young, I. T. (1977). Proof Without Prejudice: Use of the Kolmogorov-Smirnov Test for the Analysis of Histograms from Flow Systems and Other Sources. *Journal of Histochemistry & Cytochemistry, 25*(7), 935–941.

INDEX

© The Editor(s) (if applicable) and The Author(s) 2019
S. Koehler, *Lobbying, Political Uncertainty and Policy Outcomes*,
https://doi.org/10.1007/978-3-319-97055-4

Printed by Printforce, the Netherlands